Lecture Notes in Economics and Mathematical Systems

Managing Editors: M. Beckmann and W. Krelle

338

James K. Ho
R.P. Sundarraj

DECOMP:
an Implementation of
Dantzig-Wolfe Decomposition
for Linear Programming

Springer Science+Business Media, LLC

Managing Editors

Prof. Dr. M. Beckmann
Brown University
Providence, RI 02912, USA

Prof. Dr. W. Krelle
Institut für Gesellschafts- und Wirtschaftswissenschaften
der Universität Bonn
Adenauerallee 24–42, D-5300 Bonn, FRG

Authors

Professor James K. Ho
College of Business Administration, Management Science Program
The University of Tennessee
Knoxville, TN 37996-0562, USA

Professor Rangaraja P. Sundarraj
Graduate School of Management
Clark University
Worcester, MA 01610-1477, USA

ISBN 978-0-387-97154-4   ISBN 978-1-4684-9397-9 (eBook)
DOI 10.1007/978-1-4684-9397-9

Preface

The decomposition approach to solving large, complex problems plays an important role in exploiting parallel computation made possible by the latest development in computer architecture. The orignal problem is divided into a number of smaller, independent subproblems whose solutions, when suitably coordinated, produce the desired result. The coordinating procedure is usually iterative in nature. At each iteration, the independent subproblems can be solved concurrently on parallel processors. Apart from computational efficiency, this approach has significant interpretation as a model for the competitive equilibrium of decentralized systems. Since in reality, information among components of a system is processed concurrently, decomposition methods using parallel computation can provide insight into the dynamics of such interaction.

For linear optimization models that can be formulated as linear programs with the block-angular structure, i.e. independent subproblems with coupling constraints, the Dantzig-Wolfe decomposition principle provides an elegant framework of solution algorithms as well as economic interpretation. This monograph is the complete documentation of DECOMP: a robust implementation of the Dantzig-Wolfe decomposition method in FORTRAN. The code can serve as a very convenient starting point for further investigation, both computational and economic, of parallelism in large-scale systems. It can also be used as supplemental material in a second course in linear programming, computational mathematical programming, or large-scale systems.

The code had evolved over a period of more than fifteen years. Many researchers played significant roles in its development, the foremost being John Tomlin, Carlos Winkler, Etienne Loute and Benoit Culot. Funding for related projects that helped sustained DECOMP had been provided by the Department of Energy, the Office of Naval Research, the European Economic Community, and the Belgian Ministry of Scientific Policy. The preparation of this monograph is partially supported by the Office of Naval Research under Grant N00014-89-J-1528.

The entire text and all ilustrations were prepared by the authors using MacWrite and MacDraw on Apple Macintosh computers with an Apple Laser printer.

J. K. H.
R. P. S.
Knoxville, June 1989

Table of Contents

CHAPTER 1   INTRODUCTION

    1.1   Overview   1

    1.2   Scope and Purpose   3

    1.3   Availability of the DECOMP Code   4

CHAPTER 2   SPECIFICATIONS FOR A ROBUST CODE

    2.1   The Revised Simplex Method   5

    2.2   Specifications for a Robust Implementation of RSM   6

    2.3   Pseudo-Code for RSM   11

    2.4   Numerical Example of RSM   13

    2.5   Dantzig-Wolfe Decomposition   19

    2.6   Specifications for a Robust D-W Code   26

    2.7   Pseudo-Code for DECOMP   30

CHAPTER 3   PROGRAM SUBROUTINES

    3.1   Subroutine BTRAN   33

    3.2   Subroutine CHANGE   36

    3.3   Subroutine CHECK   39

    3.4   Subroutine CHSOL   49

    3.5   Subroutine CHUZR   53

    3.6   Subroutine FORMC   61

    3.7   Subroutine FTRAN   66

    3.8   Subroutine INDATA   73

    3.9   Subroutine INPUT   83

    3.10   Subroutine INVERT   91

    3.11   Subroutine ITEROP   110

    3.12   Subroutine MASTER   112

    3.13   Subroutine NORMAL   120

    3.14   Subroutine PACK   131

    3.15   Subroutine POLICY   141

    3.16   Subroutine PRICE   145

    3.17   Subroutine RESULT   149

	3.18	Subroutine UNPACK	157
	3.19	Subroutine UNRAVL	158
	3.20	Subroutine UPBETA	163
	3.21	Subroutine VECTOR	164
	3.22	Subroutine WRETA	165

CHAPTER 4		PORTABILITY ISSUES	
	4.1	Direct Access Device	168
	4.2	NAMELIST Statements	170

CHAPTER 5		USER'S GUIDE	
	5.1	Dimension of Arrays	172
	5.2	Input Data	173
	5.3	Output Data	175
	5.4	An Example	179
	5.5	Variable Dictionary	196

| BIBLIOGRAPHY | 203 |
| INDEX | 205 |

CHAPTER 1

Introduction

1.1. Overview

DECOMP is a Fortran code of the Dantzig-Wolfe (D-W) decomposition algorithm for solving block-angular linear programs. Originally coded in 1973 by Carlos Winkler at the Systems Optimization Laboratory (SOL) at Stanford University, DECOMP was built around John Tomlin's LPM1 (Tomlin [1973]), an all-in-core implementation of the revised simplex method. Since then James Ho and his European collaborators, notably Etienne Loute at the Center for Operations Research and Econometrics in Belgium, had expanded and improved upon the code as well as adapting it to run on various machines, including IBM's 370 series, CDC's Cyber series and DATA General's MV8000. It was the prototype for subsequent implementations based on commercial software (e.g. DECOMPSX with IBM's MPSX/370 in Ho & Loute [1981]) that provided significant benchmark results in LP decomposition (Ho & Loute [1983]). More recently, R.P. Sundarraj adapted the code for DEC's VAX computers in both the UNIX and VMS environment. DECOMP, as documented herein, is dimensioned to solve problems with up to 4000 rows, 10,000 columns and 55,000 non-zero elements and is intended primarily to be an experimental tool for research on computational aspects of large scale linear programming. Also, it has proven to be robust and relatively portable and may actually be useful for routine applications in certain computing environments.

Since its introduction by George Dantzig and Philip Wolfe in 1960, the decomposition approach to large, structured linear programs has only met with limited success in practical applications. Early attempts indicated that convergence may be poor. Later on, because of tremendous advances in sparse matrix techniques for the revised simplex method, it became even more difficult to compete directly with commercial LP software. This is especially true on problems that can be routinely processed by the latter. For an overview of the historical development in an updated perspective, the reader is referred to Ho [1987].

One major feature of decomposition is the decoupling of the subproblems. This independence of the component problems lends itself naturally to parallel processing. However, it is only recently, with the advent of multi-processor computers, that the potential advantages of D-W decomposition algorithms can be empirically explored. Initial results are very promising. In Ho et al [1988], an implementation known as DECOMPAR was described. It ran on an experimental multicomputer, the CRYSTAL system (Dewitt et al [1987]) which consisted of twenty VAX 11/750 minicomputers connected in a token ring architecture at the University of Wisconsin at Madison. This software tool was effective in several research projects before the CRYSTAL system became obsolete and went out of commission. One of the projects involved the demonstration of parallel decomposition applied to planning models in electric power generation (EPRI [1989]). Test cases were derived from electric generation dispatch, which seeks to operate a collection of generating units to meet power demands under various technological and regulatory constraints; and in multiregional electric generation expansion, which selects regional capacity expansions that allow the system to meet demands through inter-regional power exchange. Parallel decomposition was shown to be a viable approach. Another project led to new insights into the dynamics of information in distributed decision systems (Ho & Lee [1989]). In the interpretation of D-W decomposition (See Dantzig [1963], Burton & Obel [1977], [1980], Dirickx & Jennergren [1979]) as the decentralized coordination of coupled, semi-autonomous subsystems, the effect of timing of communication among the agents had heretofore not been closely examined. In an effort to explain the performance of DECOMPAR on various classes of problems, the concept of information schemes was developed and their dynamics were shown to play a major role in the behavior of the system.

The relatively short life span of an experimental system such as CRYSTAL is actually witness to the rapidity with which commercial multicomputer technology matures and penetrates the market. By 1988, no less than a dozen machines, of both distributed and shared memory architecture are available and becoming increasingly cost-effective. An implementation of D-W decomposition known as DECUBE has recently been completed for an Intel iPSC/2-d6, a hypercube computer with 64 processors (Ho & Gnanendran [1989]). It is shown that material requirements planning LP's (Ho & McKenney [1988]) with 30,000 constraints can be solved in about ten minutes. Previously, such performance would require a mainframe computer costing at least twenty times as much.

Both DECOMPAR and DECUBE are based on the serial code DECOMP. With the continuing trend in parallel computing with multiple processors (see e.g. Hillis [1987]), there is no doubt that significant interest and further research in LP decomposition will be renewed. As robust experimental codes for large-scale optimization are non-trivial to assemble, DECOMP can be used as a powerful building block. While this code went through many stages of development, there has never been a comprehensive documentation. This monograph presents a detailed documentation of DECOMP and serves as a programmer's guide to both its design and coding.

1.2. Scope and Purpose

DECOMP is not a commercial production code. However, in order to become an effective and robust experimental tool, it has evolved into a rather sophisticated and complex program. Our purpose is to incorporate as much tutorial material as appropriate into an otherwise purely technical documentation. Most readers will only be interested in how such a code works. For them, Chapter 2 reviews the D-W decomposition principle as well as specifications for an efficient implementation. A pseudo-code for the algorithm is also presented. The brief description of the major subroutines in Chapter 3 may also be of interest. Actual users of the code will need the user's guide in Chapter 5. It includes instructions to prepare the input data, and interpretation of the output data. A complete run of a small example problem is also given. Serious students of the implementation of a LP decomposition code and programmers who need to modify, adapt or extend the code are referred to the detailed comments in Chapter 3, the portability issues in Chapter 4 and the dictionary of variables in Chapter 5.

In should be remarked that the underlying LP solver in DECOMP is an efficient all-in-core revised simplex code for sparse linear programs based on J. Tomlin's LPM1 (Tomlin [1973]). The heart of this solver is a pre-assigned pivot scheme for the L-U factorization of the basis matrix. Although general treatments can be found in modern LP textbooks (e.g. Murtagh [1981], Chvátal [1983]), details of such implementation are scarce in the literature. The documentation of the revised simplex (FORMC, BTRAN, PRICE, FTRAN, CHUZR, etc), and basis factorization (INVERT) subroutines in Chapter 3 can therefore also serve to fill this gap.

The original Dantzig-Wolfe decomposition principle (Dantzig & Wolfe [1960]) provides a general framework for an entire class of convergent algorithms. For experimental purposes, many strategic options are built into DECOMP. These tend to complicate any effort to explain the code because of the multitude of cases that arise. For this reason, the reader should keep in mind that substantial cross referencing may be necessary to track down certain fine points and that it is impossible to provide (at least not in a linear text) a comprehensive guide to all paths of significant interest. With knowledge of the decomposition method in Chapter 2, the reader can use the pseudo code for DECOMP at the end of that chapter as the "big picture" and progressively fill in details from Chapter 3 as warranted by individual interest.

While DECOMP is a fully functional code that has proved to be robust for a diverse collection of test problems, there may still be remaining programming errors or flaws in the logical design. Also, no specific effort has gone into cleaning up the code in terms of programming style and conventions.

1.3 Availability of the DECOMP Code

DECOMP is in the public domain and has been distributed to a number of researchers worldwide over the years. It is available on a diskette subject to a nominal material and handling fee. Inquiries should be addressed to Professor J.K. Ho, Management Science Program, University of Tennessee, Knoxville, TN 37996, USA.

CHAPTER 2

Specifications For A Robust Code

2.1. The Revised Simplex Method

For completeness and ease of reference the revised simplex method (RSM) is first described briefly. Consider the linear program (LP):

	Minimize	$c\,x$	(2.1)
(LP1)	subject to	$A\,x\ =\ b$	(2.2)
		$x\ \geq\ 0$	(2.3)

where A is m by n, c is 1 by n and b is m by 1. Let

c_B be the vector formed from the elements of c corresponding to the basic variables;

$c1_B$ be the basic cost vector in Phase 1;

c_j be the jth element of the vector c;

$c1_j$ be the jth element of the Phase 1 objective vector;

a_j be the jth column of the matrix A;

B be the basic matrix (or the basis);

d_j be the vector obtained by updating a_j with the basis B;

π be the dual vector in Phase 2;

$\pi1$ be the dual vector in Phase 1; and

x^* be the solution vector.

An LP in which the constraints in (2.2) are equalities is said to be in standard form. Any LP can be transformed to the standard form by introducing slack variables, surplus variables and artificial variables depending on the type of the constraint (less than or greater than or equality). These intuitive terms are cumbersome when it comes to explaining the implementation. Hence all variables that are introduced to bring the problem to standard form will be referred to as *logical variables* or simply as *logicals*. We therefore assume that there is a unit column for a logical variable for each row in A. For feasibility, the logical must be zero for an equality row; nonnegative for a less-than-or-equal-to row; and nonpositive for a greater-than-

or-equal-to row. The revised simplex algorithm can be stated in the following steps.

1. Form the basic cost vector. In Phase 2 it is c_B and in Phase 1 it is $c1_B$.

2. Compute the dual vector. Solve $\pi B = c_B$ for π in Phase 2 and $\pi 1 B = c1_B$ for $\pi 1$ in Phase 1.

3. Compute the reduced cost for all columns j that are non-basic. In Phase 1 the reduced cost, $c1_j - \pi 1 a_j$ is simply $- \pi 1 a_j$ since $c1_j$ is zero if the jth column is non-basic. In Phase 2 the reduced cost is $c_j - \pi a_j$.

4. Find the entering column based on the reduced cost. If there is no entering column then declare the problem optimal if in Phase 2 or infeasible if in Phase 1 and stop.

5. Solve $Bd_j = a_j$ to update the entering column a_j to d_j .

6. Find the pivot row using the ratio test. If none then declare the problem unbounded and stop.

7. Return to step 1.

2.2. Specifications for a Robust Implementation of RSM

The following techniques are used in an efficient implementation of RSM.

2.2.1. Column-wise Packing

In order to exploit sparsity of the matrix A only non-zero elements are stored. The data structure consists of three arrays as described below.

(a) Array A stores the non-zero elements of the data matrix.

(b) Array IA stores the row indices of the non zero elements.

(c) Array LA stores the position in A of the first non-zero element in each column .

Therefore, the number of non-zero elements in the ith column is given by LA(i+1) - LA(i). Whenever a column of the data matrix is needed during the course of the simplex method, it has to be constructed using the information in the above arrays. This is known as unpacking and is done as follows. Suppose the ith column is to be unpacked into the array Y.

1. Initialize Y to the zero array.

2. Find the number of non-zero elements in the ith column by computing LA(i+1) - LA(i).

3. Start with the LA(i)th element of A. Use IA to assign the element to the appropriate component of the array Y and repeat this procedure for all non-zero elements of the ith column.

For example, consider the data matrix

$$\begin{bmatrix} 1 & 0 & 0 \\ 0 & 3 & 6 \\ 2 & 5 & 0 \end{bmatrix}$$

Then

$$\begin{aligned} A &= [1, 2, 3, 5, 6] \\ IA &= [1, 3, 2, 3, 2] \\ LA &= [1, 3, 5, 6] \end{aligned}$$

To unpack the 1st column consisting of 2 (= LA(2) - LA(1)) elements, Y is first initialized to the zero vector. Then A(1) is assigned to Y(1) since IA(1) is 1 and A(2) is assigned to Y(3) since IA(2) is 3.

2.2.2 *Factorization of the Basis*

Consider an n by n basis matrix B. Suppose B_1 is obtained from B by replacing its rth column with a column vector s. Then

$$B_1 = B + (s - B\, u_r)\, u_r^t$$

where u_r is the rth unit column vector and u_r^t is its transpose. Therefore

$$\begin{aligned} B_1 &= B\,[\,I + B^{-1}\,(s - Bu_r)u_r^t\,] \\ &= B\,[\,u_1,...,u_{r-1}, t, u_{r+1},...,u_n] \\ &= B\,F_1 \end{aligned}$$

where

$$t = B^{-1}s. \tag{2.6}$$

If the operation was carried out k times then

$$B_k = B_0 F_1 ... F_k \tag{2.7}$$

With this factorization of the basis, steps 2 and 5 of RSM are executed as follows.

Step 2 : To solve for π at the kth iteration of RSM we have

$$\pi B_k \quad = \quad c_B$$

which after using (2.7) becomes

$$\pi \quad = \quad (((\, c_B E_k\,)\, E_{k-1})...)\, E_1 \qquad (2.8)$$

where

$$E_i \quad = \quad F_i^{-1} \qquad \text{for all } i = 1,...,k.$$

Since F_i is an elementary matrix, E_i is also an elementary matrix. Therefore the solution for π at the kth step involves k inner products. Each of the elementary matrix E_i, i = 1 to k, is known as an *eta matrix*. The non-trivial column of an eta matrix is called an *eta vector* and the set of all eta vectors generated is commonly called the *eta file*.

Step 5 : To solve for d_j at the kth iteration of RSM we have

$$B_k d_j \quad = \quad a_j$$

which after using (2.7) becomes

$$d_j \quad = \quad E_1\, (...(\, E_{k-1}\, (\, E_k a_j\,))). \qquad (2.9)$$

2.2.3. Augmented Basis

The cost row is treated as a non-binding row in the original set of constraints, thereby changing the size of the basis to (m+1) by (m+1). Therefore the LP in (2.1)-(2.3) is rewritten as shown below

$$
\begin{array}{lllllll}
& \text{Minimize} & -s_0 & & & & (2.10) \\
& \text{subject to} & s_0 & + & cx & = & 0 & (2.11) \\
\text{(LP2)} & & & & Ax & = & b & (2.12) \\
& & & & x & \geq & 0 & (2.13)
\end{array}
$$

Note that the jth column of LP1 would correspond to the $(j+1)^{st}$ column of LP2. Since the logical variable s_0 is unrestricted in sign (2.11) is effectively a non-binding constraint. Also, if s_0 starts off as a basic variable it will never leave the basis. Let

c be the vector $[1, c]$;

x be the vector $[s_0, x]$;

A be the matrix

$$\begin{bmatrix} s_0 & c \\ 0 & A \end{bmatrix}$$

c_B be the basic cost vector of the augmented system;

c_j be the jth element of c;

a_j be the jth column of A;

B be the basic matrix of the augmented system;

d_j be the column obtained by updating a_j with the basis B;

π be the dual vector of the augmented system,

x^* be the solution vector of the augmented system.

We restate various steps of RSM in terms of LP2.

Step 1 : In Phase 2 the basic cost vector is

$$c_B \quad = \quad [-1, \mathbf{0}] \tag{2.15}$$

In Phase 1 the basic cost vector has in the ith position a 1 if the ith logical is above its upper bound; a -1 if the ith logical is below its lower bound logical; a 0 otherwise.

This scheme has the following advantages

(a) Any row can be used as the objective row by selecting its logical to be optimized.

(b) Phase 1 and Phase 2 differ only in the definition of c_B.

Step 2 : In the augmented system the new basis and its inverse are given by

$$\mathbf{B} = \begin{bmatrix} 1 & c_B \\ 0 & B \end{bmatrix} \tag{2.17}$$

$$\mathbf{B}^{-1} = \begin{bmatrix} 1 & -\pi \\ 0 & B^{-1} \end{bmatrix} \tag{2.18}$$

In Phase 1

$$\pi = [\, 0 , -cl_B \,] \, \mathbf{B}^{-1} \qquad = \quad [\, 0, -\pi l\,]. \tag{2.19}$$

In Phase 2

$$\pi = [\, 1, 0 \,] \, \mathbf{B}^{-1} = \quad [\, 1, -\pi\,]. \tag{2.20}$$

Hence, in both cases the dual vector of the original problem (LP1) is, modulo a minus sign, the second component of the dual vector of the augmented system.

Step 3 : A column a_{j+1} of the augmented system is given by

$$a_{j+1} \quad = \quad [c_j, a_j]^t \tag{2.21}$$

The jth column of LP1 corresponds to the $(j+1)^{st}$ column of LP2. Therefore, the reduced cost of the jth column of LP1 is calculated by pre-multiplying the dual vector of the augmented system to a_{j+1}. In Phase 1

$$\pi a_{j+1} \quad = \quad [0, -\pi I] \; [c_j, a_j]^t \quad = \quad -\pi I \; a_j; \tag{2.22}$$

and in Phase 2

$$\pi a_{j+1} \quad = \quad [1, -\pi] \; [c_j, a_j]^t \quad = \quad c_j - \pi a_j \tag{2.23}$$

which are the appropriate reduced costs in each case.

Step 5 : The column update of the augmented system is

$$d_{j+1} = \begin{bmatrix} 1 & -\pi \\ 0 & B^{-1} \end{bmatrix} \begin{bmatrix} c_j \\ a_j \end{bmatrix} = \begin{bmatrix} c_j - \pi a_j \\ B^{-1} a_j \end{bmatrix} \tag{2.24}$$

The second term of d_{j+1} gives the required column update.

2.2.4. Multiple Pricing

In RSM, the reduced costs of all the non-basic columns are compared and the one with the "best" reduced cost is normally chosen as the entering column. Usually, the dual prices at an iteration would not differ greatly from those of the previous iteration, and a number of columns which priced out as candidates in the previous iteration would remain candidates with respect to the new prices. The following multiple pricing scheme is used to take advantage of the above observation in order to economize on the computation intensive simplex pricing step.

(a) Instead of considering only the column with the "best" reduced cost, a pool of candidate columns (sometimes called potential entering columns) with good reduced costs are selected.

(b) Every column in the pool is updated by the current basis.

(c) The column which provides the most improvement in the objective is selected to enter the basis.

(d) The basis is updated; a new eta is generated and the entering column is removed from the candidate pool.

(e) To test if the remaining columns in the pool qualify as candidates with respect to the new basis, the new reduced costs are computed as follows. First pre-multiply the column updates by the new eta. Then pre-multiply by the appropriate cost vector.

(f) Steps (c) through (e) are repeated until all columns in the pool have either entered the basis or do not qualify as candidates.

2.3. Pseudo-code for RSM

The pseudo-code of the components of DECOMP that implements the revised simplex algorithm is given below.

Program SIMPLEX ;

Procedure *INVERT* ;
- Performs L-U factorization of the basis and also updates the solution.

Procedure *FORMC* ;
- Checks feasibility of current solution to determine Phase 1 or Phase 2 basic cost vector.

Procedure *BTRAN* ;
- Computes the dual price vector.

Procedure *PRICE* ;
- Computes the reduced costs of non-basic columns.
- Selects a pool of candidate columns.

Procedure *FTRAN* ;
- Updates a column by the current basis.
- Computes the reduced cost if called with a suitable flag during a multiple pricing pass.

Procedure *UNPACK* ;
- Unpacks a column stored in column-wise packing scheme.

Procedure *CHUZR* ;

      - Finds the pivot row for a given entering column.

      - Signals unboundedness if no pivot is found.

Procedure *UPBETA* ;

      - Updates the solution after a basis change.

Procedure *WRETA* ;

      - Writes an eta to the eta file.

Begin   /* Driver Routine */

      - Read the LP ;

      - Initialize the basis;

      -*INVERT* ;

      -*FORMC* ;

      -*BTRAN* ;

      -*PRICE* ;

      While there are candidate columns   /* multiple pricing*/

            -*UNPACK* all candidate columns;

          Repeat

               For each candidate column

                   If the column has not yet entered basis

                       - *FTRAN* ;

                       - If the reduced cost is "good" *CHUZR* ;

                       - Update the "best" improvement in the objective.

                   End If

               End For

            - Select "best" column, if any, as entering column ;

            - If entering column is selected

                - *UPBETA* ;

                - Update the basis.

                -*WRETA* ;

             End If

          Until no entering column;

          - *INVERT* if the number of etas is too large or if the inversion frequency has been reached.

$$- FORMC \; ;$$
$$- BTRAN \; ;$$
$$- PRICE \; ;$$

End While

End

Note that a few changes are necessary to incorporate RSM into the Dantzig-Wolfe decomposition algorithm discussed below in §§ 2.5 through 2.7. The details of these changes are explained in the section for subroutine NORMAL in Chapter 3.

2.4. Numerical Example of RSM

Consider the LP

Minimize $\quad - 2x1 - 3x2 - 4x3$

subject to

$x1 + x2 + x3$		$\leq$	50
$2x1 + x2 + 3x3$		$\geq$	60
$x1 - x2 + 2x3$		$=$	20
$x1, \; x2, \; x3$		$\geq$	0

After input of the data the LP is converted to

Minimize $\quad -s0$

subject to

$s0$	$-2x1$	$-3x2$	$-4x3$	$= 0$
$s1$	$+x1$	$+x2$	$+x3$	$= 50$
$-s2$	$+2x1$	$+x2$	$+3x3$	$= 60$
$s3$	$+x1$	$-x2$	$+2x3$	$= 20$

$s1, s2, \; x1, \; x2, \; x3 \geq 0; \; s3 = 0$

Iteration 1 :

Basic set = $\{s0, s1, s2, s3\}$;   Non-basic set = $\{x1, x2, x3\}$

Step 1 : Call FORMC to form the cost vector. Since s2 and s3 violate their bounds, we are in Phase 1. The initial solution, the basic cost vector, the initial basis and its inverse are :

$$\mathbf{x}^* \quad = \quad [0, 50, -60, 20]^t \qquad (2.25)$$
$$\mathbf{c_B} \quad = \quad [0, 0, 1, -1]$$

$$\mathbf{B}_0 \;=\; \begin{bmatrix} 1 & & & \\ & 1 & & \\ & & -1 & \\ & & & 1 \end{bmatrix} \;=\; \mathbf{F}_0 \qquad\qquad (2.26)$$

$$\mathbf{B}_0^{-1} \;=\; \mathbf{B}_0 \;=\; \mathbf{E}_0 \qquad\qquad (2.27)$$

It should be noted that the basis is factorized and represented by a series of LU etas. It is never actually inverted. This is explained in the section on subroutine INVERT in Chapter 3. We use the inverse in this example and in later sections to simplify the exposition.

Step 2 : Call BTRAN to compute the dual prices. Using $\mathbf{c}_B$ and $\mathbf{B}_0$ found above we get

$$\pi \;=\; [0, 0, -1, -1].$$

Step 3 : Call PRICE to compute the reduced cost of all non-basic columns and also to select a pool of candidate columns with "good" reduced costs. The reduced costs are

$$[0, 0, -1, -1] \begin{bmatrix} -2 & -3 & -4 \\ 1 & 1 & 1 \\ 2 & 1 & 3 \\ 1 & -1 & 2 \end{bmatrix} \;=\; [-3, 0, -5]$$

The pool of candidate columns consists of x1 and x3.

Step 3a : *Multiple Pricing*

      Pass 1

      (a) Call FTRAN to update the first column in the pool (x1) using
$$\mathbf{d}_j = \mathbf{B}_0^{-1}\,\mathbf{a}_j.$$
This results in

$$\mathbf{d}_j = \begin{bmatrix} 1 & & & \\ & 1 & & \\ & & -1 & \\ & & & 1 \end{bmatrix} \begin{bmatrix} -2 \\ 1 \\ 2 \\ 1 \end{bmatrix} = \begin{bmatrix} -2 \\ 1 \\ -2 \\ 1 \end{bmatrix}$$

(b) Call CHUZR to find the pivot row for this column. The ratios are :

$$\{\ -\ ,\ 50/1,\ -\ ,\ 20/1\ \}$$

The minimum ratio is given by row 4. The change in objective value is

$$(-3)*(20)\ =\ -60.$$

(c) The "best" change so far is -60 with a pivot in the column corresponding to $x1$ and row 4.

(d) Call FTRAN to update the next column in the pool ($x3$). The updated column is

$$\mathbf{d}_j\ =\ [-4, 1, -3, 2]^t$$

(e) Call CHUZR to find the pivot row for $x3$. The ratios are :

$$\{\ -\ ,\ 50/1,\ -\ ,\ 20/2\ \}$$

The minimum ratio is given by row 4. The change in the objective value is $(-5)(10)\ =\ -50.$

(f) The "best" change is still given by step c. Therefore, the entering variable is $x1$ and the pivot row is row 4.

(g) Call UPBETA to update the solution. This results in

$$\mathbf{x}^*\ =\ [0+40, 50-20, -60+40, 20]^t$$
$$=\ [40, 30, -20, 20]^t \tag{2.28}$$

(h) We then have

Basic set = $\{s0, s1, s2, x1\}$; Non-basic set = $\{x2, x3, s3\}$.

A flag is set so that on the next call FTRAN would also compute the reduced cost.

Pass 2 : The basis and its inverse at this point are

$$\mathbf{B}_1 = \begin{bmatrix} 1 & & & \\ & 1 & & \\ & & -1 & \\ & & & 1 \end{bmatrix} \begin{bmatrix} 1 & & & -2 \\ & 1 & & 1 \\ & & 1 & -2 \\ & & & 1 \end{bmatrix} \tag{2.29}$$

$$=\qquad \mathbf{F}_0 \qquad\quad \mathbf{F}_1 \tag{2.30}$$

$$B_1^{-1} = \begin{bmatrix} 1 & & & 2 \\ & 1 & & -1 \\ & & -1 & 2 \\ & & & 1 \end{bmatrix} \begin{bmatrix} 1 & & & \\ & 1 & & \\ & -1 & & \\ & & & 1 \end{bmatrix} \tag{2.31}$$

(a) Call FTRAN to update the column corresponding to x3 and also to compute the reduced cost. The new update

$$\mathbf{d'_j} = \mathbf{E_1 d_j} = \mathbf{E_1} \ [-4, 1, -3, 2]^t = [0, -1, 1, 2]^t$$

where $\mathbf{d_j}$ is the previous update. The reduced cost

$$\pi \mathbf{a_j} = \mathbf{c_B B_1^{-1} a_j} = [0, 0, 1, 0] \ [0, -1, 1, 2]^t = 1.$$

Since this is positive the column corresponding to x3 can no longer be a candidate to enter and other columns should be priced out using the latest dual vector.

Iteration 2 :

Basic set = {s0, s1, s2, x1};   Non-basic set = {x2, x3, s3}.

Current basic values = $[40, 30, -20, 20]^t$.

Step 1 : Call FORMC to form the basic cost vector. Reference to (2.28) gives

$$\mathbf{c_B} = [0, 0, 1, 0].$$

Step 2 : Call BTRAN to compute π. Reference to (2.31) and (2.32) gives

$$\pi = [0, 0, 1, 0] \ \mathbf{E_1 \ E_0} = [0, 0, -1, 2].$$

Step 3 : Call PRICE to compute the reduced cost of all non-basic columns to select a pool of candidate columns with "good" reduced costs. The reduced costs are

$$[\ 0, 0, -1, 2\ ] \begin{bmatrix} -3 & -4 \\ 1 & 1 \\ 1 & 3 \\ -1 & 2 \end{bmatrix} = [\ -3, 1\ ]$$

The pool consists of one candidate column corresponds to x2.

Step 3a : Multiple Pricing. Since the pool has only one column there is only one multiple pricing pass.

> *Pass 1 :*
>
> (a) Call FTRAN to update the column corresponding to x2. The update is

$$\mathbf{d}_j = \mathbf{E}_1 \mathbf{E}_0 \ [-3, 1, 1, -1]^t = [-5, 2, -3, -1]^t$$

> (b) Call CHUZR to find the pivot row. The ratio is {30/2} and is given by row 2 which becomes the pivot row.
>
> (c) The change in the objective value is $(-3)(15) = -45$ with a pivot in the column corresponding to x2 and row 2. Since there is only one column in the pool, the entering variable is x2 and the leaving variable corresponds to row 2.
>
> (d) Call UPBETA to update the solution. The updated solution is

$$\begin{aligned}
\mathbf{x}^* &= [40+5(15), 15, -20+3(15), 20+1(15)]^t \\
&= [115, 15, 25, 35]^t
\end{aligned} \tag{2.33}$$

Iteration 3 :

> Basics set= {so, x2, s2, x1}; Non-basics set= {x3, s3, s1}.

Step 1 : Call FORMC to form the basic cost vector. Since all basic variables are within bounds, we are in Phase 2, and

$$\mathbf{c}_B \quad = \quad [-1, 0, 0, 0].$$

The basis and its inverse are

$$\mathbf{B}_2 = \mathbf{F}_0 \ \mathbf{F}_1 \begin{bmatrix} 1 & -5 & & \\ & 2 & & \\ & -3 & 1 & \\ & -1 & & 1 \end{bmatrix} \tag{2.34}$$

$$= \quad \mathbf{F}_0 \ \mathbf{F}_1 \ \mathbf{F}_2 \tag{2.35}$$

$$B_2^{-1} = \begin{bmatrix} 1 & 2.5 & & \\ & 0.5 & & \\ & 1.5 & 1 & \\ & 0.5 & & 1 \end{bmatrix} E_1 \, E_0 \tag{2.36}$$

$$= \quad E_2 \, E_1 \, E_0 \tag{2.37}$$

Step 2 : Call BTRAN to compute π. Reference to (2.36) and (2.37) gives

$$\pi \quad = \quad [1, 0, 0, 0] \ E_2 \, E_1 \, E_0 \quad = \quad [1, 2.5, 0, -0.5].$$

Step 3 : Call PRICE to compute the reduced costs of all non-basic columns to select a pool of candidate columns with "good" reduced costs. The reduced costs are

$$[1, 2.5, 0, -0.5] \begin{bmatrix} 0 & -4 \\ 1 & 1 \\ 0 & 3 \\ 0 & 2 \end{bmatrix} = \quad [2.5, -2.5]$$

The pool consisting of one candidate column corresponding to x3 which is $[-4, 1, 3, 2]^t$.

Step 3a : *Multiple Pricing*. Since there is only candidate column in the pool, there is only one multiple pricing pass.

> *Pass 1* :
>
> (a) Call FTRAN to update the column corresponding to x3. The update is
>
> $$d_j \quad = \quad E_2 \, E_1 \, E_0 \ [-4, 1, 3, 2]^t \ = \ [2.5, -0.5, -0.5, 1.5]^t$$
>
> (b) Call CHUZR to find the pivot row. The ratios are
> $$\{ \, - \, , \, - \, , \, - \, , \, 35/1.5 \, \}.$$
>
> (c) The change in the objective value is $(-2.5)*(23.33)$ with a pivot in row 4.
>
> (d) Call UPBETA to update the solution. The solution is

$$x^* = [115+2.5(23.33), \; 15+.5(23.33), \; 25+1.5(23.33), \; 23.33]^t$$
$$= [173, \; 26.6, \; 36.6, \; 23.3]^t.$$

Iteration 4 :

Basic set = $\{s0, x2, s2, x3\}$;   Non-basic set = $\{x1, s3, s1\}$.

After calls to FORMC, BTRAN and PRICE it can be seen that there will be no entering column and this establishes optimality.

2.5. Dantzig-Wolfe Decomposition

2.5.1. Theorem on Convex Combinations

In order to present an overview of the Dantzig-Wolfe (D-W) decomposition a theorem on convex combinations is stated below without proof.

Theorem : Let $X = \{x \mid Ax = b, \; x \geq 0\}$. Then a point x is in X if and only if it can be written as a convex combination of its extreme points plus a nonnegative linear combination of generators of extreme rays (homogeneous solutions) of X; i.e.

$$x = \sum_j (\beta_j \, x^j) \tag{2.38}$$

where

$$\sum_j (\beta_j \partial_j) = \quad 1 \quad ; \quad \beta_j \geq 0 \;\; \text{for all } j \tag{2.39}$$

$$\partial_j \quad = 0 \text{ if } x^j \text{ is an extreme ray generator of } X \tag{2.40}$$
$$1 \text{ if } x^j \text{ is an extreme point of } X.$$

2.5.2 Development of the Decomposition Principle

Consider the LP

(LP3) $A_1 x = b_1$ (2.42)

                                     $A_2 x = b_2$ (2.43)

                                     $x \geq 0$ (2.44)

where A_1 is m_0 by n, A_2 is m by n and all other vectors and matrices havie suitable dimensions. Let

$$S = \{x \mid A_2 x = b_2, \ x \geq 0 \}$$ (2.45)

Then by the theorem on convex combinations any point in S is given by

$$x = \sum_j (\beta_j x^j)$$ (2.46)

where

$$\sum_j (\beta_j \partial_j) = 1 \quad ; \quad \beta_j \geq 0 \ \text{for all j and}$$ (2.47)

$\partial_j = 0$ if x^j is an extreme ray generator of S;
         1 if x^j is an extreme point of S.   (2.48)

The index j runs over all extreme points and extreme ray generators of S. Substituting for x in LP3 gives

Minimize $\sum_j (c x^j)\beta_j$ (2.49)

subject to

(LP4) $\sum_j (A_1 x^j)\beta_j = b_1$ (2.50)

                $\sum_j (\beta_j \partial_j) = 1$ (2.51)

                $\beta \geq 0$ (2.52)

with ∂_j defined as in (2.48).

Problems LP3 and LP4 are equivalent since, among all solutions of (2.43) only those that also satisfy (2.42) are chosen by LP4. LP4 is called the master problem and (2.51) - (2.52) of the master problem are known as the convexity constraints. The master problem has $(m_0 + 1)$ rows compared to the $(m_0 + m)$ rows of the original problem LP3. On the other hand it has as many columns as there are extreme points and rays in S and this number can be astronomical.

Moreover, these columns are not known a-priori. An elegant scheme developed by Dantzig and Wolfe [1960] to find these columns is described below.

Column generation : To solve LP4 with RSM, all the steps described in §2.1, except the one of finding the reduced cost, can be done routinely. To find the reduced cost

$$\mathbf{cx}^j - \pi \begin{bmatrix} \mathbf{A}_1 \ \mathbf{x}^j \\ \partial_j \end{bmatrix} \tag{2.53}$$

the column $[\ \mathbf{A}_1\mathbf{x}^j, \partial_j\ ]^t$ is required but is not readily available. However, the pricing procedure can be done implicity as follows. Partitioning

$$\pi = [\pi_0, \mu] \tag{2.54}$$

where π_0 is the dual vector of the constraints in (2.50) and μ is the dual vector of the constraint (2.51) the reduced cost is

$$(\mathbf{c} - \pi_0\mathbf{A}_1)\ \mathbf{x}^j - \mu\partial_j. \tag{2.55}$$

To find the entering column by the usual column selection rule (i.e. most negative reduced cost) we need

$$\min_j \{(\mathbf{c} - \pi_0\mathbf{A}_1)\mathbf{x}^j - \mu\partial_j \} \tag{2.56}$$

This is equivalent to solving the following LP:

$$\begin{array}{lll} \text{Minimize} & (\mathbf{c} - \pi_0\mathbf{A}_1)\mathbf{x} \\ \text{subject to} \\ \text{(LP5)} & \mathbf{A}_2\mathbf{x} & = & \mathbf{b}_2 \\ & \mathbf{x} & \geq & \mathbf{0}. \end{array}$$

LP5 is called the *subproblem*. Three cases may result.

  (a) The subproblem yields an optimal solution $\mathbf{x}^*$ such that

$$(\mathbf{c} - \pi_0\mathbf{A}_1)\mathbf{x}^* > \mu\partial_j \tag{2.57}$$

in which case there are no entering columns in the master problem LP4, which in turn implies that LP4 is either optimal or infeasible.

  (b) The subproblem yields an optimal solution $\mathbf{x}^*$ and it passes the *candidacy test* given by (2.58).

$$(c - \pi_0 A_1)x^* < \mu \partial_j \tag{2.58}$$

The required entering column is given by (2.59).

$$\begin{bmatrix} cx^* \\ A_1 x^* \\ 1 \end{bmatrix} \tag{2.59}$$

(c) The subproblem is unbounded from below in which case a homogeneous solution x^* that satisfies $(c - \pi_0 A_1)x^* < 0$ can be obtained. The entering column in LP4 is given by

$$\begin{bmatrix} cx^* \\ A_1 x^* \\ 0 \end{bmatrix} \tag{2.60}$$

These entering columns in the master problem generated by solving the subproblem are called *proposals*.

2.5.3 D-W Algorithm for Block-Angular Problems

A block-angular LP with R blocks has the form:

Minimize $\qquad z = \sum (c_r x_r) \tag{2.61}$

subject to

(LP6)

$$\sum (A1_r x_r) = b_0 \tag{2.62}$$

$$A2_r x_r = b_r \qquad r = 1,\ldots, R \tag{2.63}$$

$$x_r \geq 0 \qquad r = 0,\ldots, R \tag{2.64}$$

where c_r is 1 by n_r, b_r is m_r by 1 and all other vectors and matrices are of suitable dimensions. Schematically a block angular problem is shown in Figure 2.1.

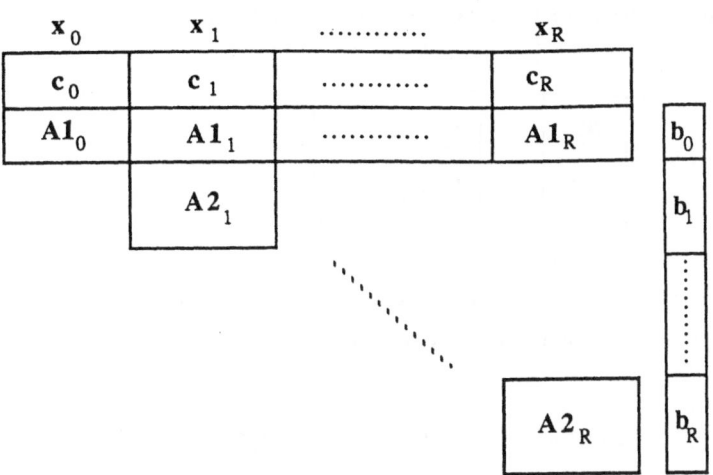

Figure 2.1 A block-angular linear program.

Assuming that each block of constraints (2.62) and (2.63) is feasible, the m_0 constraints in (2.62), called the coupling constraints, will be used to define a master problem which, as seen later, differs from LP4 in two respects:

(a) it has R convexity rows, one for each block or sub-system; and

(b) each block is represented by a separate set of ß's.

At cycle k of the algorithm, the master problem determines prices π_0^k on the constraints (2.62) and (2.63). With these prices, a subproblem from block r at cycle k can be defined by

$$\text{Minimize} \qquad v_r^k \quad = \quad (c_r - \pi_0^k A1_r)x_r \qquad (2.65)$$

subject to

(LP7_r^k)
$$A2_r x_r \quad = \quad b_r \qquad (2.66)$$

$$x_r \quad \geq \quad 0 \qquad (2.67)$$

(LP7_r^k) has either an extreme point or an extreme ray solution. In either case, if the solution x_r^{*k} passes the candidacy test, the k_r^{th} proposal generated by block r in cycle k for the master problem can be defined as

$$\begin{bmatrix} p_r^{kr} \\ q_r^{kr} \end{bmatrix} = \begin{bmatrix} c_r x_r^{*k} \\ A1_r x_r^{*k} \end{bmatrix} \qquad (2.68)$$

24

Representing all the proposals from block r as a matrix, we have

$$
\begin{bmatrix} \mathbf{p}_r^{kr} \\ \mathbf{Q}_r^{kr} \end{bmatrix} = \begin{bmatrix} \mathbf{p}_r^1, \ldots, \mathbf{p}_r^{kr} \\ \mathbf{q}_r^1, \ldots, \mathbf{q}_r^{kr} \end{bmatrix} \tag{2.69}
$$

The master problem (LP8^k) shown below is then obtained from (2.61) and (2.62) by substituting $\mathbf{x}_r$, $r = 1,\ldots R$, with the convex combinations of the proposals.

$$
\begin{aligned}
\text{Minimize} \quad & z^k = \mathbf{c}_0 \mathbf{x}_0 + \Sigma(\mathbf{p}_r^{kr} \ß_r^{kr}) && (2.70) \\
\text{subject to} & \\
\text{(LP8}^k\text{)} \quad & \mathbf{A1}_0 \mathbf{x}_0 + \Sigma(\mathbf{Q}_r^{kr}ß_r^{kr}) = \mathbf{b}_0 && (2.71) \\
& \partial_r^{kr}ß_r^{kr} = 1 && (2.72) \\
& \mathbf{x}_0 \geq \mathbf{0}, \quad ß_r^{kr} \geq \mathbf{0}; \quad r = 1,\ldots R && (2.73)
\end{aligned}
$$

where $ß_r^{kr} = [ß_r^1,\ldots, ß_r^{kr}]^t$ with $ß_r^l$ being the weight of the lth proposal from block r; and $\partial_r^{kr} = [\partial_r^1,\ldots, \partial_r^{kr}]$ with ∂_r^l being one if the lth proposal corresponds to an extreme point and zero otherwise. Finally, denote the dual vector of (LP8^k) by $[\pi_0^k, \mu_1^k,\ldots, \mu_R^k]$ where π_0^k corresponds to the coupling constraints (2.71) and μ_r^k to the rth convexity constraint in (2.72). The algorithm, consisting of three phases can now be described.

Phase 1 (For feasibility) :

Step 1 : (Feasibility of each block). Solve subproblem (LP7$_r^k$) for $r = 1$ to R. If one of the subproblems is infeasible, then the whole problem is infeasible; hence STOP. Otherwise, generate a proposal for the master problem (LP8^1).

Step 2 : (Feasibility of coupling constraints). Use Phase 2 to minimize the sum of infeasibilities in (LP8^1). If the optimal value is positive then the problem is infeasible; hence STOP. Otherwise, restore the original objective function and proceed to Phase 2.

Phase 2 (For optimality) :

Step 0 : Set k to 1 if the objective function is a sum of infeasibilities. If the objective is the original one, then set z'^{k-1} as the lower bound on the optimal value of the objective function.

Step 1 : Solve the current master problem (LP8^k). If it is unbounded, then the original problem is unbounded; hence STOP. Otherwise, send the dual vector $[\pi_0^k, \mu_r^k]$ to subproblem r for r = 1 to R.

Step 2 : Solve subproblem (LP7$_r^k$). Use the candidacy test given by (2.74) to determine whether a proposal can be generated according to (2.68) for the master problem.

$$v_r^{*k} \quad = \quad (c_r - \pi_0^k A1_r)x^*_r{}^k \quad < \quad \mu_r^k \qquad (2.74)$$

Step 3 : (Stopping criterion). If no proposal has been generated in Step 2 of Phase 2 then STOP. If the objective function is the sum of infeasibilities, terminate Step 2 of Phase 1. Otherwise proceed to Phase 3. If all subproblems have been optimized and the objective function is the original one, update the lower bound on the optimal value according to:

$$z'^k \quad = \quad \max(z'^{k-1},\ z^k + \sum (v_r^{*k} - \mu_r^k)) \qquad (2.75)$$

If $z^k - z'^k < \Delta$, where Δ is some small, positive, user supplied tolerance, then proceed to Phase 3. Otherwise set k to k+1 and go to Step 1 of Phase 2.

Phase 3 (For reconstruction of the primal solution):

Step 1 (Compute allocation). Let the optimal solution of (LP8^k) be $[z^{*k}; x_0^*, \beta_1^{*k},..., \beta_R^{*k}]$ Compute

$$y_r = Q_r^{kr}\beta_r^{*kr} \quad , \text{ for } r = 1,..., R \text{ and} \qquad (2.76)$$

define for r = 1,...R:

	minimize	$c_r x_r$		
	subject to			
(LP9$_r$)		$A1_r x_r =$	y_r	(2.77)
		$A2_r x_r =$	b_r	
		$x_r \geq$	0	

Step 2 (Reconstruction). Solve (LP9$_r$) for x_r^*, r = 1,..., R. Then $[x_0^*, x_1^*,..., x_R^*]$ constitutes a solution to the original problem with objective value z^{*k}.

2.6. Specifications for a Robust D-W Code

There are three major aspects in the efficient design of a decomposition code :

a. Solving the master and subproblems with RSM. This aspect has already been discussed in § 2.2.

b. Data handling to update and solve the master and subproblems. This aspect is discussed in § 2.6.1.

c. Computational strategies discussed in § 2.6.2.

2.6.1. Augmenting Coupling Constraints to a Subproblem

As seen in § 2.5.3, the objective function in a subproblem given by (2.65) changes from cycle to cycle. However, with the column packing scheme for sparse matrices, it is not convenient to compute and store the objective function explicitly which would involve the insertion and deletion of nonzero elements. Moreover, generating a proposal by multiplying x_r explicitly by $A1_r$ would require a distinction among the $A1_r$ and $A2_r$ data stored compactly in the proposed data structure. Both of these difficulties can be overcome with the following method.

Recall that in RSM the objective function is required only for finding the reduced cost. This can be done if the subproblem is organized as shown in Figure 2.2. The basis B_r, and its inverse B_r^{-1}, corresponding to this structure are shown in Figures 2.3 and 2.4 respectively. The steps of RSM and D-W decomposition affected by this structure are now stated as follows.

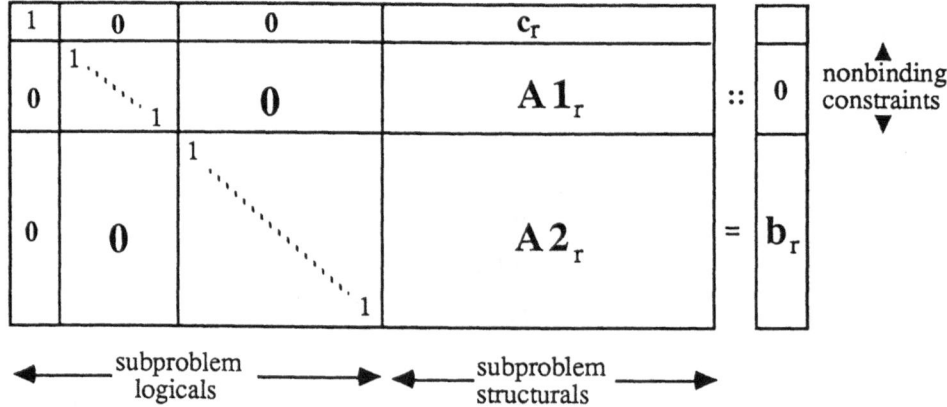

Figure 2.2 Organization of a subproblem

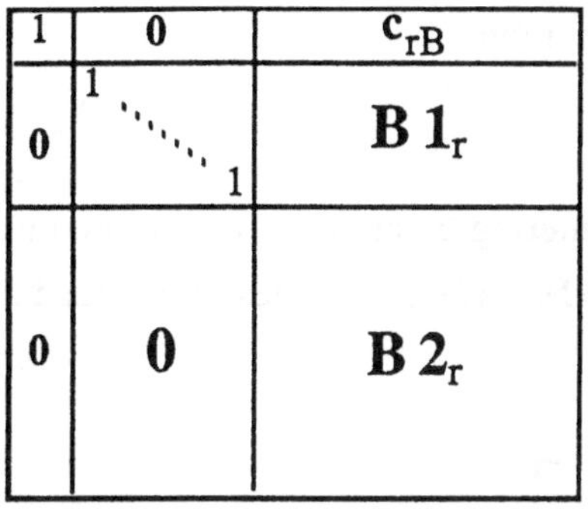

1	0	c_{rB}
0	$\begin{smallmatrix}1 & & \\ & \ddots & \\ & & 1\end{smallmatrix}$	$\mathbf{B1}_r$
0	0	$\mathbf{B2}_r$

1	0	$-c_{rB}\mathbf{B2}_r^{-1}$
0	$\begin{smallmatrix}1 & & \\ & \ddots & \\ & & 1\end{smallmatrix}$	$-\mathbf{B1}_r\mathbf{B2}_r^{-1}$
0	0	$\mathbf{B2}_r^{-1}$

Figure 2.3 A subproblem basis. 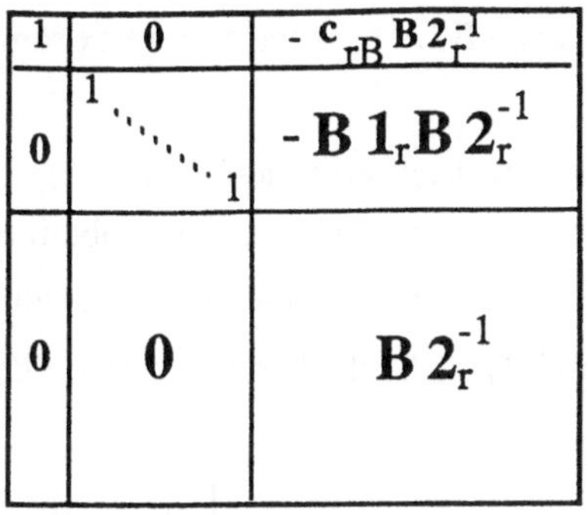   Figure 2.4 A subproblem basis inverse.

Step 1 of RSM : (Basic cost vector). The basic cost vector in a subproblem is taken to be

$$[1, -\pi_0^k, 0] \tag{2.78}$$

Step 2 of RSM : (Dual prices) Pre-multiplying (2.78) to the inverse given in Figure 2.4 yields :

$$[1, -\pi_0^k, -\pi_r^k] \tag{2.79}$$

so that the dual vector is given, modulo a minus sign, by the third term.

Step 3 of RSM : (Reduced costs) Pre-multiplying (2.79) to the jth non-basic column of the matrix given by Figure 2.2 yields $c_j - \pi_0^k \mathbf{A1}_{rj} - \pi_r^k \mathbf{A2}_{rj}$. This is by definition the required reduced cost.

Extreme point proposal generation.

Pre-multiplying the basis inverse by the right hand side vector $[0, 0, \mathbf{b}_r]^t$, yields

$$\begin{bmatrix} -c_{rB}\ \mathbf{B2}_r^{-1}\ \mathbf{b}_r \\ -\mathbf{B1}_r\ \mathbf{B2}_r^{-1}\ \mathbf{b}_r \\ \mathbf{B2}_r^{-1}\ \mathbf{b}_r \end{bmatrix} \tag{2.80}$$

in which the first two components contains the required extreme point proposal defined in (2.68) since $B2_r^{-1}b_r$ is the current extreme point solution.

Extreme ray proposal generation.

An extreme ray solution is obtained if an entering column is found to be unbounded. Suppose column j, which can be written as $[c_{rj}, A1^t_{rj}, A2^t_{rj}]^t$ implies unboundedness, then pre-multiplying by the basis inverse yields:

$$\begin{bmatrix} c_{rj} & -c_{r_B}B2_r^{-1}A2_{rj} \\ A1_{rj} & -B1_r B2_r^{-1}A2_{rj} \\ & B2_r^{-1}A2_{rj} \end{bmatrix} \tag{2.81}$$

An extreme ray (a homogeneous solution to (2.63) and (2.64)) is constructed by setting

$$y_j = 1;$$
$$y_s = -u_i[B2_r^{-1}A2_{rj}] \text{ if } x_s \text{ is basic in row i; and}$$
$$y_s = 0, \text{ otherwise.}$$

u_i is the ith unit row vector. Using this solution in (2.68) we see that the required extreme ray proposal is given by the first two component of (2.81).

Together, (2.80) and (2.81) imply that both extreme point and extreme ray proposals can be obtained from data maintained by RSM for the subproblem as organized above. In the first case, the proposal (modulo a sign) is found in the updated rhs. In the second case, it is found in the updated candidate column.

2.6.2 Computational Strategies

These are refinements of the basic algorithm aimed at accelerating convergence. They are by and large heuristic in nature and their benefits cannot be guaranteed on theoretical grounds.

Intermediate stage proposal generation.

Since any solution of the subproblem passing the candidacy test given by (2.74) can lead to an improvement in the master problem, it can be used to generate a proposal. This allows the possibility of obtaining a multitude of proposals from a subproblem during a single cycle. Empirical experience (Ho and Loute [1981]) showed that this strategy can accelerate convergence significantly. An example of a mechanism to control proposal generation depends on three factors :

a. frequency control;

b. relative improvement of the reduced cost of the potential proposal; and

c. total number of proposals generated.

The user specifies four parameters : q_{freq}, q_{perc}, q_{max} and q_{intr}. A proposal is generated as soon as the current solution passes the candidacy test. Then a proposal is generated every q_{freq} iterations or whenever the reduced cost of the potential proposal is improved by q_{perc}. No more than q_{max} proposals are sent to the master. The current cycle is interrupted once a total of q_{intr} proposals have been generated by all subproblems during the current cycle.

Interrupted cycles.

The convergence of the decomposition algorithm is not lost if a cycle is interrupted. In fact we will see how a cycle interruption can be beneficial in some cases. A cycle can be interrupted for any of the following reasons :

1. As seen above, a cycle is interrupted if during the course of solving a subproblem, a total of q_{intr} proposals have been generated. In the next cycle, we start with the next subproblem. This is done to give all subproblems an equal opportunity to generate proposals.

2. In contrast to the above reasoning, very often it may be useful to favor some subproblems to generate more proposals than others. This is done if some knowledge of the nature of the problem indicates that some of the subproblems are likely to be more active in proposal generation than others. In this case a cycle is interrupted if a subproblem is not optimal and has generated q_{max} proposals. The next cycle is started with the same subproblem.

Terminal stage proposal generation.

A subproblem is in its terminal stage in a cycle if it is optimal or if an unbounded column is discovered. If the subproblem is optimal, one proposal is generated corresponding to the extreme point of the optimal solution. When it is unbounded, two companion proposals are generated. The first proposal is from the generator of the extreme ray that caused the unboundedness and the second one is from the extreme point corresponding to the current basic feasible solution. In order to discover other extreme rays, the column causing undoundedness is fixed at zero and the optimization of the restricted subproblem is continued.

Proposal purging.

As the size of the master problem grows with the addition of new proposals, the inactive columns in it must be purged periodically to keep storage requirements within acceptable limits.

2.7. Pseudo-code for DECOMP

The pseudo-code of the program DECOMP, a FORTRAN version which incorporates the robust specifications of § 2.6 into the block-angular D-W algorithm developed in § 2.5.3, is described below.

Program *DECOMP* ;

/*Only those procedures which are critical to explaining DECOMP are shown. Procedures with names in italics represent actual subroutines in DECOMP. Others are logical units in the code that are named here only to simplify the presentation. */

Procedure *INPUT* ;

       - Read MPS data for the current problem.

Procedure *CHECK* ;

       - Check to see if a proposal can be written to the proposal buffer.

Procedure *PACK* ;

       - Pack proposals from a proposal buffer into the data array.

Procedure SIMPLEX ;

/* Refer to § 2.3 for details */

       - Solve the given LP.

Procedure *INVERT* ;

       - Do L-U factorization of the given basis and update the
       solution.

Procedure *NORMAL* ;

       - *INVERT* if eta size gets too large or if inversion frequency is
       reached;
       - *PACK* if solving master problem and if there are candidates;
       - SIMPLEX ;

- *CHECK* if SIMPLEX identifies an optimal solution.

Procedure *INDATA* ;
/* Performs Step 1 of D-W Phase 1 */
   For each problem
     - *INPUT* ;
     If the problem is not the master problem
      - *NORMAL* ;
      If the problem is infeasible then ;
       - Declare the whole problem infeasible and STOP.
      Else
       - Write the proposal to its buffer.
      End If
     End If
   End For
   - Read the master problem into core ;
   -*PACK.*

Procedure *POLICY(KK)* ;
   - Based on the strategy, set KK to the number of problems that have to be solved in the current cycle. Also, set the order in which the problems are to be solved and the proposal policy.

Procedure SOLVMSTR(KCYC, RELDIFF) ;
/* To optimize master problem */
   - Increment KCYC to indicate the beginning of a new cycle ;
   - *NORMAL* ;
   - If the master problem (coupling constraints) is infeasible then STOP;
   - Compute the relative difference in the primal and dual objective value (RELDIFF).
   - If RELDIFF > tolerance for the stopping criterion (RSTOP) then
     - Save the dual vector of the master problem in a buffer.
   End If

Procedure SOLVSUB(KKK, KCYC) ;
/* To optimize subproblem */

- *NORMAL* ;

- Decide if the current cycle is to be interrupted. (cf. Section 2.6.2)

- If cycle is to be interrupted then

        - SOLVMSTR(KCYC, RELDIFF) ;

        - If RELDIFF > RSTOP then the first subproblem to be solved in the new cycle depends on the user's strategy.

                Case 1 : the KKK^{th} subproblem is the first subproblem in the new cycle. In this case SOLVSUB(KKK, KCYC) ;

                Case 2 : the $(KKK+1)^{st}$ subproblem is the first subproblem of the new cycle. In this case just proceed.

        End If

Procedure *MASTER* ;

/*Performs Step 2 of D-W Phase 1 and D-W Phase 2 */

        - Set the cycle number KCYC to 0 ;

        - SOLVMSTR(KCYC, RELDIFF) ;

        - While the RELDIFF is greater than tolerance of the stopping criterion

            - *POLICY(KK)* ;

            - For each problem KKK = 1 to KK-1

                - SOLVSUB(KKK, KCYC) ;

            End For

            - SOLVMSTR(KCYC, RELDIFF) ;

      End While

Procedure *RESULT* ;

/*Performs D-W Phase 3 */

        - Compute the allocation to each subproblem;

        - Solve subproblem ($LP9_r$), r = 1 to R, to reconstruct the solution $x_r{}^*$.

Begin   /* Main Routine */

        - *INDATA* ;

        - *MASTER* ;

        - *RESULT* ;

End

CHAPTER 3

Program Subroutines

In this chapter the functions of all subroutines in the program are explained in detail in alphabetical order. The explanation is organized under a combination of the following headings: *Major Task, Inputs, Outputs, Called By, Calling, Description, Steps* and *Detailed Comments* (in certain cases simply *Comments*). The *Description* section contains a narrative of the functions of the subroutine, a summary of which is given in the *Steps* section. The actual code is related to these functions in *Comments*. The nature of the contents under the other headings is self explanatory.

3.1. Subroutine BTRAN

3.1.1. Major Tasks

Compute the simplex multipliers π by solving the system $\pi B = c_B$.

3.1.2. Inputs

1. The buffer YA(*, *) with c_B stored in one of its columns.

2. KVEC indicating the column of YA(*, *) containing c_B.

3. Eta file in the array A(*).

3.1.3. Outputs

The dual vector π in the buffer YA(*, KVEC).

3.1.4. Called By

NORMAL after the cost vector has been formed by a call to FORMC.

3.1.5. Description

The solution of π from the system $\pi B = c_B$ is obtained through transformation of c_B by a set of elementary eta matrices as indicated in (3.1)

$$\pi = c_B E_k ... E_1 \tag{3.1}$$

where

$$B = F_1 ... F_k \tag{3.2}$$

and

$$F_j^{-1} = E_j \tag{3.3}$$

Let E_j be the jth eta matrix and let the pth column of E_j be the non-trivial column as shown below.

$$E_j = [u_1,..., u_{p-1}, n_p, u_{p+1},..., u_n] \quad j = 1,..., k \tag{3.4}$$

where

$$n_p = \begin{array}{ll} -e_{ip}/e_{pp}; & i \neq p \\ 1/e_{pp}; & i = p \end{array} \tag{3.5}$$

and u_i is the ith unit vector.

Without loss of generality consider the transformation by the eta matrix E_j. Note that at this point c_B has already been transformed to c'_B (say) by eta matrices $E_k,..., E_{j+1}$. Therefore, the jth stage of the transformation is

$$c'_B E_j = (c'_B)_i \qquad\qquad i \neq p \tag{3.6}$$

$$\left\{ (c'_B)_p - \sum_{j \neq p}(c'_B)_j \, e_{jp} \right\}/e_{pp} \qquad i = p$$

$$= (c'_B)_i \qquad\qquad i \neq p \tag{3.7}$$
$$(R - S) / e_{pp} \qquad\qquad i = p$$

where

$$R = (c'_B)_p \text{ and } S = \sum_{j \neq p}(c'_B)_j \, e_{jp} \tag{3.8}$$

since only the pth column of E_j is non-trivial. From (3.6) it is clear that at the jth stage, ($j = 1,..., k$) the only element that changes in c'_B is the one corresponding to the pivot position of E_j and this element is computed by an inner product. Since the pivot element of each eta is stored first, the pivot position of the eta is readily inferred.

3.1.6. Detailed Comments

```
C*******************************************************************
C                                                            BTRAN
C    NETA    :  Number of etas generated.
C    YA(*, *):  Buffer, one of whose columns contains the cost vector.
C    KVEC    :  Indicates the column of YA(*, *) containing the cost vector.
C    KFASE   :  Normally set to 2 in which case the cost vector is set up in YA(*, 6).
C               When a subproblem's solution is infeasible in D-W Phase 2 then
C               KFASE is set to 1 and the cost vector is formed in YA(*, 5).
C    LL      :  First non-zero element of the non-trivial column of the eta $E_j$ being used in
C               the transformation.
C    KK      :  Last non-zero element of the non-trivial column of the  eta $E_j$.
C    IPIV    :  Pivot position of the non-trivial column of $E_j$.
C    A(*)    :  Array containing the eta file.
C
C    1. If there are no etas then RETURN.  Else proceed. [line 1]
C    2. Find the column in YA(*, *) where the cost vector is stored.  This is decided by
C       KFASE. [lines 3 - 4]
C    3. For each eta vector j do :
C
C                  -  Set IK to the eta index starting from the last since by (3.1) the transformation
C                     starts with the last eta and ends with the first eta.  [line 6]
C                  -  Set LL and KK to the first and the last non-zero element of the non-trivial
C                      column of $E_j$.  [lines 7 - 8].
C                  -  Since the pivot element of the eta vector is stored first, the row index of the
C                     first element of the eta vector gives the pivot position.  Set DP to pivot
C                     element. Hence, DP holds $1/e_{pp}$ of (3.6).  [lines 9 - 10]
C                  -  Set DY to R of equation (3.8).  [line 11]
C                  -  If pivot element is the only element of the eta vector then S of (3.8) is 0.
C                     In this case simply find DY*DP = R/epp and assign it to YA(IPIV, KVEC).
C                     [lines 13, 19]
C                     Else proceed.
C                  -  Find DSUM = S of (3.8). [lines 15 - 18] Also find $(R - S)/e_{pp}$ and assign it
C                     to YA(IPIV, KVEC). [line 19]
C
C       End For
```

```
C
C*******************************************************************
C                                                              BTRAN
         IF (NETA .LE. 0) GOTO 9000                              1
         CALL TICTAC(JTIM)                                       2
         KVEC = 6                                                3
         IF (KFASE .EQ. 1) KFASE = 5                             4
         DO 1000 I = 1, NETA                                     5
         IK = NETA - I ÷ 1                                       6
         LL = LE(IK)                                             7
         KK = LE(IK + 1) - 1                                     8
         IPIV = IA(LL)                                           9
         DP = A(LL)                                             10
         DY = YA(IPIV, KVEC)                                    11
         DSUM = 0.0                                             12
         IF (LL .LE. KK) GOTO 600                               13
         LL = LL + 1                                            14
         DO 500 J = LL, KK                                      15
         IR = IA(J)                                             16
         DSUM = DSUM + A(J)*YA(IPIV, KVEC)                      17
500      CONTINUE                                               18
600      YA(IPIV, KVEC) = (DY - DSUM)*DP                        19
1000     CONTINUE                                               20
         CALL TICTAC(ITINV)                                     21
         JTINV = JTINV + ITINV - JTIM                           22
         RETURN                                                 23
         END                                                    24
```

3.2. Subroutine CHANGE

3.2.1. Major Task

Exchange problems between the direct access device (DAD) and core.

3.2.2. Input

1. LSUP indicating the problem that is to be read into core memory.

2. NOLA(*, *) indicating where in the direct access device each problem is stored.

3.2.3. Called By

INDATA, NORMAL, RESULT.

3.2.4. Calling

VECTOR to actually effect the exchange of problems.

3.2.5. Description

CHANGE first checks to see if the problem to be read is already in core. If the problem is not in core then problems have to be exchanged between the DAD and core as follows. First, the position in DAD where the problem in core is to be written is obtained. The subroutine VECTOR is called to write the problem in core to DAD. Next, the position where the problem to be read is residing in DAD is obtained and again VECTOR is called to read the problem into core.

3.2.6. Steps

1. Write the problem in core to DAD.
2. Read the problem required in core from DAD.

3.2.7. Detailed Comments

```
C*************************************************************************
C     Step 1 :                                                CHANGE
C
C     LSUP       : Index of the problem to be read.
C     LCORE      : Indicates if the problem to be read is in core.
C     A(*)       : Data array.
C     NOLA(2, *): Indicates in the data array A(*) the end of the eta file of the *th problem.
C     NOLA(3, *): Indicates the position in DAD where all the data (with eta) of the *th
C                  problem is residing.
C     KIKA       : Flag to VECTOR. When one it indicates that VECTOR should write the
C                  problem currently in core. When two it indicates that the LSUPth problem
C                  is to be read from DAD by VECTOR.
C     LMAX       : Number of subproblems.
C     LMAXP1     : LMAX plus one.
C     KK         : Index of the problem that is to be read. KK = LMAXP1 indicates the master
C                  problem.
C
```

```
C
C      1. If the problem is already in core then there is nothing to be done.  Hence RETURN.
C          [line 3].  Otherwise proceed.
C      2. Set NETA to zero if the number of iterations since basis inversion has  exceeded the
C          inversion frequency or if there is not enough space in the data array A(*).
C          [lines 6 - 10]. Note that NETA = 0 triggers a re-inversion.
C      3. Set IAUX to NOLA(2, KK).  Since the eta vectors are written after the matrix data this
C          indicates the number of elements of the array A(*) that are to be read.  [lines 11, 14].
C      4. Set KD to point to the record in DAD to which the problem in core is to be written.
C          [line 13].
C      5. Set KIKA to 1 (indicating write) and call VECTOR to write the problem in core to the
C          KD^{th} record of DAD.
C
C*********************************************************************************
C                                                          CHANGE
          DIMENSION NB(3612)                                  1
          EQUIVALENCE (JH(1), NB(1))                          2
          IF(LSUP .EQ. LCORE) GO TO 5000                      3
          CALL TICTAC(LI)                                     4
          IF (LCORE .EQ. -1) GO TO 3000                       5
          INVF = INVFRQ                                       6
          IF (LCORE .EQ. 0) INVF = INVFMA                     7
          KK = LCORE                                          8
          IF(KK .EQ. 0)KK = LMAXP1                            9
          IF(NELEM.GT.(NEMAX-NRMAX).OR.ITSINV.GT.            10
         1(4*INVF) / 5)NETA = 0
          NOLA(2, KK) = LE(NETA + 1) - 1                     11
          KIKA = 1                                           12
 2050     KD = NOLA(3, KK)                                   13
          IAUX = NOLA(2, KK)                                 14
          CALL VECTOR(KIKA, A, IA, NB, IAUX)                 15


C*********************************************************************************
C      Step 2 :
C
C      Refer to step 1 for variable definitions.
C
```

C   1. Set KIKA to two to indicate read. [line 20]

C   2. Set KD to the record in DAD where the LSUP[th] problem is to be read. [lines 13]

C   3. Set IAUX to indicate the number of elements of the data array A(*) that are to be read.

C   [line 14]

C   4. Call VECTOR to read the problem into core. [line 15]

C

C**

C   CHANGE

2500   IF(LCORE .EQ. LSUP) GO TO 4000   16

3000   LCORE = LSUP   17

        KK = LSUP   18

        IF(LSUP .EQ. 0) KK = LMAXP1   19

        KIKA = 2   20

        GO TO 2050   21

4000   CALL TICTAC(LF)   22

        NX=NX + LF - LI   23

5000   RETURN   24

        END   25

3.3. Subroutine CHECK

3.3.1. Major Tasks

Checks if a proposal can be made.

3.3.2. Input

YA(*, KVEC) contains the unbounded column if any.

3.3.3. Outputs

1. When appropriate, proposals are made in D-W Phase 2. The proposals, if any, will be stored in Z(*, *).
2. Sets flag (IRTN) to indicate if an INVERT was done.
3. Sets flag KOUT to indicate that multiple pricing has to be interrupted.

3.3.4. Called By

NORMAL.

3.3.5. Calling

1. INVERT to make a more accurate proposal.
2. FORMC subsequent to the call to INVERT.
3. If an unbounded proposal is to be made, then UNPACK is called to get the column causing the unboundedness.
4. If an unbounded proposal is to be made, then FTRAN is called to update the column causing unboundedness.

3.3.6. Description

The object of CHECK is to see if a proposal can be generated from the subproblem being solved. Reference to § 2.5.3 shows that a proposal can be generated as soon as the current solution passes the candidacy test. All subsequent solutions until optimality of the subproblem can be used to generate proposals if the user chooses the appropriate strategy. Strategies can be classified into two major types. The first type permits intermediate proposal generation before optimality of the subproblem. The second type generates proposals only at optimality or unboundedness. For either type a proposal can be generated only if further criteria in addition to the candidacy test are met.

Case 1 : (No intermediate proposals) The test in this case is simply to check for optimality or unboundedness. If the subproblem is optimal then the proposal is from the extreme point corresponding to the optimal solution. If unbounded then two companion proposals are generated. One is from the column that caused the unboundedness and the other is from the extreme point corresponding to the current basic feasible solution.

Case 2 : (Intermediate proposals) There are several tests as described below.

Test A : If the reduced cost of the subproblem's entering variable is "better" than some user specified value RCTEST or if the allowable increase in the potential entering column is less than 10, then Test A is passed.

Test B : If the allowable increase in the subproblem's potential entering column is less than 100 then Test B is passed.

Percentage test : If the relative improvement in the reduced cost of the proposal column is greater than some user specified value RPERC, then the Percentage test is passed.

Frequency test : If the frequency of proposal generation exceeds some user specified value NCANFR then the frequency test is passed.

Proposals generated in this case are extreme point proposals corresponding to the current basic feasible solution.

Actual generation of proposals depends on whether an extreme point or an extreme ray is selected for generation and involves writing an extreme point solution or an extreme ray column to a column of the proposal buffer $Z(*, *)$.

Case 1 : *(Extreme point proposal)* Reference to subroutine NORMAL shows that it is possible for two consecutive calls to CHECK to correspond to the same extreme point. In this case a proposal need not be written to Z(*, *). Otherwise, the next step is to decide where in Z(*, *) to write the proposal, and this depends on the strategy chosen.

Case 1.1 : *(Intermediate proposals not permitted)* If there is space in Z(*, *) then the proposal is accommodated in the first available free column. On the other hand if the number of proposals generated exceeds the number of columns NPROS in Z(*, *), then the current contents of Z(*, *) are written to the end of the unit 8 work file and subsequent proposals are written beginning with the first column of Z(*, *).

Case 1.2 : (Intermediate proposals permitted) Let MAXCA be the number of proposals sent to the master problem. This case is further broken up into two sub-cases. In the first sub-case more than MAXCA proposals can be generated but only the last MAXCA of them are sent to the master problem. For example, assume that MAXCA proposals have been generated. Then the $(MAXCA+1)^{st}$ proposal overwrites the proposal in the first column of Z(*, *). This way, no more than MAXCA proposals are kept. Hence, the count on the number of proposals kept is incremented only if the number generated is less than MAXCA. In the second sub-case the solution of the problem is terminated for the current cycle once MAXCA proposals have been generated and the flag KOUT is set to 1. The decision about where in Z(*, *) to write the proposal is the same as in case 1.1.

The last step is to actually write the proposal to Z(*, *). This is done by copying the first NROWO (number of coupling rows) elements of the solution vector X(*) into the appropriate column of Z(*, *). Depending on the number of iterations since the last INVERT, it may be necessary to perform an INVERT before copying it to Z(*, *) to ensure numerical accuracy of the proposal.

Case 2 : *(Extreme ray proposal)* This is very similar to case 1 except for the following.

1. The proposal is found in the updated column that caused the unboundedness. It is copied to Z(*, *).

2. After writing an extreme ray column, case 1 is called to see if a companion extreme point proposal can be generated.

3.3.7. Steps

1. Determine which proposal tests, if any, have to be done.

2. Perform

    (a) Test A, Test B, Percentage Test and Frequency Test; *or*

    (b) Termination Test as is appropriate.

3. Incorporate

    (a) an extreme point proposal; *or*

(b) an extreme ray proposal and thereafter, possibly an extreme point
proposal.

3.3.8. Detailed Comments

```
C************************************************************************
C    Step 1 :                                                    CHECK
C
C    MAST   :  Indicates stage of the problem.  '0' indicates step 1 of D-W Phase 1.
C                 '1' indicates step 2 of D-W Phase 1 or D-W Phase 2. '3' indicates D-W
C                 Phase 3.
C    MSTAT  :  Indicates status of the problem.
C    DX     :  Subproblem objective value minus the dual variable of the corresponding
C                 convexity constraint.  DX is the reduced cost of the proposal column.
C    KRIT(*) :  Indicates the type of the strategy chosen by the user. '2' if intermediate
C                 proposals are allowed. '1' if proposals only at optimality of unboundedness
C                 of the subproblem.
C    LSUB   :  Indicates problem index.  '0' for master problem.
C
C    Case 1 :  (Proposal not written due to the problem's status)
C                  - While executing step 1 of D-W Phase 1 (MAST = 0) the proposal will be
C                    copied in the subroutine INDATA.  [line 4]
C                  - If the problem's current solution is infeasible or if the master problem is
C                    being solved then a proposal need not be generated.  [line 3]
C
C    Case 2 :  (Proposal not generated due to the strategy chosen by user)
C                  -  If MSTAT = QF and the user has chosen the strategy of not generating
C                    intermediate proposals then RETURN.  [lines 6 - 7]
C
C    Case 3 :  (Intermediate proposals can be generated)
C                  - Go to step 2a for further tests.
C
C    Case 4 :  (Optimal and unbounded proposal)
C                  - If MSTAT ≠ QF then it should be optimal or unbounded since if it were
C                    QI then we would have gone to case 1.  Go to step 2b
C
C************************************************************************
```

```
C                                                                    CHECK
        KOUT = 0                                                       1
        IRTN = 0                                                       2
        IF(KFASE .EQ. 1 .OR. LSUB .EQ. 0) GO TO 1000                   3
        IF(MAST .NE. 1) GO TO 1000                                     4
        DX = DOB + YTEMP(LSUB)                                         5
        IF(MSTAT .NE. QF) GO TO 2100                                   6
        MM = KRIT(KKK)                                                 7
        GO TO (5000, 2000), MM                                         8
 1000   IF(MSTAT.NE. QF .AND. MSTAT .NE. QI) KOUT = 1                  9
        IF (LSUB .NE. 0) GO TO 5000                                   10
        IF (INCO .EQ. -1) KOUT = 1                                    11
        GO TO 5000                                                    12
C***********************************************************************
C       Step 2a :
C
C       DX        : Reduced cost of the potential proposal.  [line 5]
C       DCMIN     : Reduced cost of the subproblem's candidate column.
C       RCTEST    : User supplied parameter for Test A.
C       DP        : Allowable increase in the subproblem's candidate column.
C       IPROP(*)  : Number of proposals generated under the (*) th criterion.
C       QPROP     : Criterion used for the most recently generated proposal.
C
C       - If the candidacy test is not met then RETURN.  Else proceed.  [line 13]
C       - Test A.  If Test A is not passed then proceed.   [lines 14 - 17]
C       - Test B.  If Test B is not passed then proceed.   [lines 18 - 21]
C       - Percentage Test.  If not passed then proceed. [lines 22 - 26]
C       - Frequency Test.  If not passed then RETURN.  [lines 27 - 29]
C
C***********************************************************************
 2000   IF(DX .GE. -ZTCAND) GO TO 5000                               13
        IF(DCMIN(IYIV) .GT. RCTEST .OR. DP .LT. 10.) GO TO 2010      14
        IPROP(3) = IPROP(3) + 1                                      15
        QPROP = QA                                                   16
        GO TO 2450                                                   17
 2010   IF(DP .LT. 100.) GO TO 2020                                  18
        IPROP(4) = IPROP(4) + 1                                      19
```

```
C                                                               CHECK
        QPROP = QB                                                20
        GO TO 2450                                                21
2020    DP = RPERC*DZ                                              22
        IF (DX .GE. DP) GO TO 2030                                23
        IPROP(1) = IPROP(1) + 1                                   24
        QPROP = QR                                                25
        GO TO 2400                                                26
2030    IF(ITCNT - ITSLCA .LT. NCANFR) GO TO 5000                 27
        IPROP(2) = IPROP(2) + 1                                   28
        QPROP = QF                                                29


C*******************************************************************
C       Step 2b :
C
C       - Unboundedness test.  If this test is not passed then proceed.  [lines 30 - 33]
C       - Optimality test.  [lines 34 - 39]
C       - If optimality test is passed then KOUT should be set to 1 to indicate that no more
C         simplex iterations need to be done to solve this problem.  [line 37]
C       - If the candidacy test is not passed then RETURN.  Else go to step 4 to write the
C         proposal to Z.  [line 39]
C
C*******************************************************************
2100    IF(MSTAT .NE. QU) GO TO 2110                              30
        IPROP(6) = IPROP(6) + 1                                   31
        QPROP = QU                                                32
        GO TO 3000                                                33
2110    IF(MSTAT .NE. QBL) GO TO 2400                             34
        QPROP = QO                                                35
        IPROP(5) = IPROP(5) + 1                                   36
        KOUT = 1                                                  37
2200    IF(ITCNT .EQ. ITMA) GO TO 5000                            38
        IF(DX .GE. -ZTCOST) GO TO 5000                            39
2400    IF(DABS(DX - DZ) .LE. ZTCOST) GO TO 5000                  40


C*******************************************************************
```

45

```
C       Step 3a :                                                    CHECK
C
C       MAXCA:  Number of proposals that can be sent to the master in one cycle.
C       NCASUB: Number of proposals generated by the subproblem being solved.
C       NCAND : Number of proposals kept in current cycle.
C       IPROS  : Pointer to the column in Z(*, *) where the proposal can be written.
C       NPROS  : Column dimension of Z(*, *).
C       KSTR   : User's strategy. If '5' then only the last MAXCA proposals are kept.
C                   If '1' then no intermediate stage proposal can be generated. Otherwise
C                   problem gets interrupted once MAXCA proposals are generated.
C       NROWO : Number of coupling rows.
C       X(*)   : Solution vector.  First NROWO components of X(*) contains the
C                   proposal.
C
C          - Check if the current call to CHECK corresponds to the same extreme
C            point as a previous call. [lines 43 - 44]
C          - If KSTR is 5 and MAXCA proposals have been generated then the
C            number of proposals NCASUB generated by the current subproblem
C            and the number of proposals NCAND generated in the current cycle
C            are not incremented.  This is because the proposal being generated would
C            overwrite a proposal already existing in Z(*, *). [lines 45, 49 - 52]
C          - If KSTR is 5 and MAXCA proposals have not been generated then
C            increment NCAND and NCASUB. [45 - 47]
C          - If KSTR is not 5 and more proposals are generated than can be
C            accommodated by Z(*, *), then write the current contents of Z(*, *) to the
C            unit 8 work file and write the proposal being generated to the first
C            column of Z(*, *). [lines 53 - 55]
C          - If KSTR is not 5 and there is space in Z(*, *) then write the current
C            proposal to the first available column of Z(*, *). [line 49]
C          - Check if an inversion is required and if so call INVERT and set the flag
C            IRTN to 1. [lines 56 - 59]
C          - Set KFASE to 1 and call FORMC.  If the solution is infeasible then
C            since KFASE is 1, FORMC would already have set up the cost vector in
C            the buffer YA(*, 5).  Otherwise set KFASE to 2 and call FORMC again
C            to set up the cost vector in YA(*, 6). [lines 60 - 69]
C          - Write the proposal into Z(*, *)  and the reduced cost of the proposal
C            column in the (NROWO+1)st row of Z(*, *).  [lines 70 - 72]
```

```
C                column of Z(*, *).  [lines 53 - 55]
C              - If KSTR is not 5 and there is space in Z(*, *) then write the current
C                proposal to the first available column of Z(*, *).  [line 49]
C              - Check if an inversion is required and if so call INVERT and set the flag
C                IRTN to 1.  [lines 56 - 59]
C              - Set KFASE to 1 and call FORMC.  If the solution is infeasible then
C                since KFASE is 1, FORMC would already have set up the cost vector in
C                the buffer YA(*, 5).  Otherwise set KFASE to 2 and call FORMC again
C                to set up the cost vector in YA(*, 6).  [lines 60 - 69]
C              - Write the proposal into Z(*, *)  and the reduced cost of the proposal
C                 column in the (NROWO+1)$^{st}$ row of Z(*, *).  [lines 70 - 72]
C              - Note the iteration count for proposal generation.  This is to enable the
C                first check in this step.  [line 73]
C              - If KSTR $\neq$ 5 then mark the origin of the NCAND$^{th}$ proposal.  [line 75]
C                Also check if the solution of problem is to be terminated or if the multiple
C                pricing cycle is to be interrupted. Set KOUT to 1 in both cases.
C                [lines 75, 120]
C                Otherwise do the following.  ICAND is pointer to the last proposal of
C                the previous subproblem.  Since IPROS gives the proposal position
C                beyond ICAND, mark the origin of the (ICAND + IPROS)$^{th}$ proposal.
C                [lines 77 - 78]
C
C*******************************************************************************
C                                                                      CHECK
2450    DZ = DX                                                           41
        IF (KSTR .EQ. 5 .AND. NCUNB .EQ. MAXCA) GO TO 5000                42
        IF (ITSLCA .EQ. ITCNT) GO TO 5000                                 43
        IF(MSTAT .EQ. QU .AND. ITSLCA .EQ. ITCNT - 1) GOTO 5000           44
        IF (KSTR .EQ. 5 .AND. NCASUB .GE. MAXCA) GO TO 2452               45
        NCAND = NCAND + 1                                                 46
        NCASUB = NCASUB + 1                                               47
        IF (ISUNB .GT. 0) NCUNB = NCUNB + 1                               48
2452    IPROS = IPROS + 1                                                 49
        IF (KSTR .NE. 5) GO TO 2453                                       50
        IF (IPROS .GT. MAXCA) IPROS = 1                                   51
        GO TO 2455                                                        52
2453    IF (IPROS .LE. NPROS) GO TO 2455                                  53
```

```
C                                                               CHECK
        IF (JSTAT .EQ. QBL .OR. JSTAT .EQ. QU) MSTAT = JSTAT      65
        DOB = 0                                                   66
        DO 2470 I = 1, NROWO                                      67
2470    DOB = DOB -X(I)*YA(I, 6)                                  68
        CALL FORMC                                                69
2475    DO 2500 I = 1, NROWO                                      70
2500    Z(I, IPROS) = -X(I)                                       71
        Z(NROWO+1, IPROS) = DOB + YTEMP(LSUB)                     72
        ITSLCA = ITCNT                                            73
        IF (KSTR .EQ. 5) GO TO 2550                               74
        NAME(NCAND) = LSUB                                        75
        GO TO 4000                                                76
2550    NAME(ICAND + IPROS) = LSUB                                77
        GO TO 5000                                                78
C*********************************************************************
C       Step 3b :
C
C       Same as step 4 but for two differences :
C
C       1. Call FTRAN to update the unbounded column by the current basis. [line 107]
C          Write the updated extreme ray in the buffer YA(*, *) to Z(*, *).
C          [lines 110 - 111]
C       2. If maximum limit on proposal generation has not been exceeded then
C          go to Step 3a to generate an extreme point proposal.  [lines 115 - 119]
C
C*********************************************************************
3000    KOUT = 1                                                  79
        IF (ISUNB .GT. 0) NCUNB = NCUNB + 1                       80
        IF (KSTR .EQ. 5 .AND. NCASUB .GE.MAXCA) GO TO 3010        81
        NCAND = NCAND + 1                                         82
        NCASUB = NCASUB + 1                                       83
3010    IPROS = IPROS + 1                                         84
        IF (KSTR .NE. 5) GO TO 3015                               85
        IF (IPROS .GT. MAXCA) IPROS = 1                           86
        GO TO 3050                                                87
3015    IF (IPROS .LE. NPROS) GO TO 3050                          88
```

```
C                                                                  CHECK
        WRITE (8) Z                                                  89
        IPROS = 1                                                    90
3050    IF(ITSINV .LE. INVFRQ) GO TO 3100                            91
        CALL INVERT                                                  92
        ITSINV = 0                                                   93
        IRTN = 1                                                     94
        KFASE = 1                                                    95
        CALL FORMC                                                   96
        IF (MSTAT .NE. QI) GO TO 3060                                97
        WRITE (6, 6666) LSUB                                         98
6666    FORMAT('0LOSS OF FEASIBILITY IN SUBPROBLEM ', I4)            99
C
        STOP                                                        100
3060    KFASE = 2                                                   101
        DOB = 0                                                     102
        DO 3080 I = 1, NROWO                                        103
3080    DOB = DOB  -  X(I)*YA(I, 6)                                 104
        CALL FORMC                                                  105
        MSTAT = QU                                                  106
        J = JCOLP(KVEC)                                             107
        CALL UNPACK(J, KVEC)                                        108
        CALL FTRAN(1, KVEC)                                         109
3100    DO 3200 I = 1, NROWO                                        110
3200    Z(I, IPROS) = YA(I, KVEC)                                   111
        Z(NROWO + 1, IPROS) = DCMIN(KVEC)                           112
        IF (KSTR .EQ. 5) GO TO 3250                                 113
        NAME(NCAND) = -LSUB                                         114
        IF(NCAND .LT. MAXNCA) GO TO 2200                            115
        GO TO 5000                                                  116
3250    NAME(ICAND + IPROS) = -LSUB                                 117
        IF (NCASUB .LE. NC(KKK)) GO TO 2200                         118
        GO TO 5000                                                  119
4000    IF(NCAND .GE. MAXNCA .OR. NCASUB .GE. NC(KKK))              120
       1KOUT = 1
5000    RETURN                                                      121
        END                                                         122
```

3.4. Subroutine CHSOL

3.4.1. Major Task

Check accuracy of current solution to determine if a reinversion is necessary.

3.4.2. Input

1. The rhs **b** of the LP in the buffer YA(*, 3).
2. The array X(*) containing **x**+∂**x** of (3.10).

3.4.3. Output

1. The (approximate) condition number of the basis **B** if in step 1 of D-W Phase 1.
2. The infinity norm of ∂**b** or ∂**x** (cf. 3.11 and 3.12).

3.4.4. Called By

NORMAL.

3.4.5. Calling

FTRAN.

3.4.6. Description

Let **x** and **x**+∂**x** be the true and the computed solutions of the system

$$\mathbf{B}\mathbf{x} = \mathbf{b} \tag{3.9}$$

where **B** and **b** are a basis and the rhs vector of the given LP respectively. We then have

$$\mathbf{B}(\mathbf{x}+\partial\mathbf{x}) = \mathbf{b}+\partial\mathbf{b} \tag{3.10}$$

$$\partial\mathbf{x} = \mathbf{B}^{-1}\partial\mathbf{b} \tag{3.11}$$

where ∂**b** is known as the residual rhs and is given by

$$\partial\mathbf{b} = \mathbf{B}(\mathbf{x}+\partial\mathbf{x}) - \mathbf{b} \tag{3.12}$$

The condition number K of the basis **B** is given by

$$K = K1/(MN) \tag{3.13}$$

where

$$M = (1 - \| \partial b B^{-1} \|)^{-1} \qquad (3.14)$$

$$N = \| \partial b \| / \| b \| + \| \partial B \| / \| B \| \qquad (3.15)$$

$$K1 = \| \partial x \| / \| x \| \qquad (3.16)$$

Since ∂b and ∂B are small M is close to 1 and N is small. Therefore, the approximate condition number is determined by first computing K1 and then multiplying it by a large number. Note that this approximate condition number is used for experimental purposes and is not used in the logical design of DECOMP.

3.4.7. Steps

1. Compute $\| x \|$. Use (3.12) to compute ∂b.
2. Compute $\| \partial b \|$.
3. Compute $\| \partial x \|$ and the approximate condition number if necessary.

Note : By default the norm used is the infinity norm.

3.4.8. Detailed Comments

```
C*************************************************************************
C    Step 1 :                                                      CHSOL
C
C    X(*)      : Contains the solution vector x.
C    XMAX      : At the end of step 1 it gives ||x||.
C    NROW      : Number of rows in the problem under consideration.
C    JH(*)     : Indicates the column that is basic in the *th row.
C    IV        : Indicates the basic column of the row being considered.
C    YA(*, 3)  : Before step 1 contains the rhs and at the end of step 1 it has the residual rhs.
C    A(*)      : Array containing the data of the problem.
C    IA(*)     : Array indicating the row index in the matrix of each element in the array A(*).
C
C    For each row i do :
C
C            1. Set IV to index of the basic column in the ith row.
C            2. If the ith row is non-binding then set XMAX <-- |X(i)| if
C               XMAX is less than the absolute value of X(i). [lines 3 - 6]
C            3. Subtract from the rhs (YA(*, 3)) the contribution due to the IVth column of the
C               matrix A. This is given by the formula : ∂b <--- ∂b - a_IV x_IV. [lines 8 - 12]
C
```

```
C     End For
C
C**********************************************************************
C                                                              CHSOL
      XMAX = 0.                                                   1
      DO 1000 J = 1, NROW                                         2
      IV = JH(J)                                                  3
      IF (IV .GT. NROW) GO TO 90                                  4
      IF (ISTYPE(IV) .EQ. 0) GO TO 100                            5
   90 IF (DABS(X(J)) .GT. XMAX) XMAX = DABS(X(J))                 6
  100 IF (IV .EQ. 0) GO TO 1000                                   7
      LL = LA(IV)                                                 8
      KK = LA(IV + 1) - 1                                         9
      DO 700 I = LL, KK                                          10
      IR = IA(I)                                                 11
      YA(IR, 3) = YA(IR, 3) - A(I)*X(J)                          12
  700 CONTINUE                                                   13
 1000 CONTINUE                                                   14


C**********************************************************************
C     Step 2 :
C
C     YA(*, 3): Contains the residual rhs ∂b.  If the program is executing INDATA
C               (step 1 of D-W Phase 1) then it contains ∂x as given by (3.11).
C     ERMAX: At the end of step 2 it contains ‖∂b‖.
C
C     1. For each row i set ERMAX to max{ERMAX, |YA(i,3)|}. [lines 15 - 18]
C     2. Call FTRAN to compute ∂x = B⁻¹∂b. [lines 19 - 20]
C
C**********************************************************************
      ERMAX = 0.                                                 15
      DO 2000 I = 1, NROW                                        16
      IF (DABS(YA(I, 3)) .GT. ERMAX) ERMAX = DABS(YA(I, 3))      17
 2000 CONTINUE                                                   18
      IF(MAST .GE. 1) GO TO 10000                                19
      CALL FTRAN(1, 3)                                           20
```

```
C*********************************************************************
C       Step 3 :
C
C       NROW        : Number of rows in the problem.
C       EEMAX       : At the end of step 3 it contains the maximum row error.
C       XMAX        : Conains ||x||.
C       YA(*, 3)    : At the end of step 2 it contains ∂x.
C       COND        : At the end of step 3 it contains the approximate condition number.
C       JH(*)       : Indicates the basic column of the *th row.
C       IV          : Indicates the basic column of the row being considered.
C       ISTYPE(*)   : Indicates the type (equality, etc..) of the IVth  row.
C
C       For each row i do :
C
C          1. Obtain the basic variable of the ith row.  [line 23]
C          2. If it is binding then set EEMAX <-- max{EEMAX, YA(i, 3)}. [lines 24 - 26]
C
C       End For.
C       Compute the approximate condition number.  [line 28]
C
C*********************************************************************
C                                                              CHSOL
        EEMAX = 0.                                                21
        DO 3000 I = 1, NROW                                       22
        IV = JH(I)                                                23
        IF (IV .GT. NROW) GO TO 2500                              24
        IF (ISTYPE(IV) .EQ. 0) GO TO 3000                         25
 2500   IF (DABS(YA(I, 3)) .GT. EEMAX) EEMAX = DABS(YA(I, 3))     26
 3000   CONTINUE                                                  27
        COND = (EEMAX*2**24) / XMAX                               28
        IF(ERMAX .LT. ZTOLZE) GO TO 10000                         29
        WRITE(6, 8000) ERMAX, COND                                30
 8000   FORMAT(21H0MAXIMUM ROW ERROR = , D14.5 / 36H              31
       1 APPROX CONDITION NUMBER OF BASIS = , D14.5)
10000   RETURN                                                    32
        END                                                       33
```

3.5. Subroutine CHUZR

3.5.1. Major Task

To find the pivot row. In case of ties, find the largest pivot from among those which gives the same best change in the objective value.

3.5.2. Inputs

1. The buffer YA(*, KVEC) which contains the candidate column stored in one of its columns; *or*

2. The flag (JENTRY) set to '2' to find the maximum pivot row for the pivot column. In this case the following inputs are given.

a. The pivot column in the buffer YA(*, KVEC).

b. The variable DE giving the allowable increase in the entering variable corresponding to the pivot column.

c. The variable NTYPE giving the ratio type (cf. § 3.5.5.) for the pivot column.

3.5.3. Outputs

If not flagged (JENTRY ≠ 2) then

- NTYPE gives the ratio type;
- DE gives the allowable increase in the entering variable if the variable corresponding to the candidate column enters.
- DP gives the change in the objective value if the variable corresponding to the candidate column enters.

Else

- IROWP gives row index of maximum pivot.

End if

3.5.4. Called By

NORMAL.

3.5.5 Description

If CHUZR is not called with a flag then the leaving variable is decided based on two broad objectives :

1. To drive a basic equality logical out of the basis at the first opportunity without creating any new infeasibilities.

2. In the absence of basic equality logicals, to remove as many infeasibilities as possible if there are any remaining infeasibilities.

Since the rhs is not necessarily positive, some changes as illustrated in the example below will have to be made to the normal RSM ratio test.

Example:

$x2$			$+4x1$	$=$	$+36$	(3.17)
	$x3$		$+1x1$	$=$	$+24$	(3.18)
		$x4$	$-2x1$	$=$	-36	(3.19)
			$x5$	$-3x1$	$=$	-24	(3.20)
			$a1$	$-3x1$	$=$	-36	(3.21)
				$a2$	$-9x1$	$=$	-36	(3.22)

Here $a1$ and $a2$ are equality logicals, $x2$ to $x5$ are variables that are not equality logicals and $x1$ is the entering variable. The following observations can be made with this example.

1. Unlike the normal RSM ratio test (3.19) - (3.22) cannot be ignored. In fact in keeping with the first objective of driving the artificials out of the basis, the leaving variable is $a2$. Note however, that $a1$ could not be driven out of the basis since this would make $x2$ infeasible. Also note that the leaving variable $a2$ corresponds to the min ratio of (3.21) and (3.22).

2. If the coefficient of $x1$ in (3.22) is -1 instead of -9 then no equality logical can be driven out of the basis without creating new infeasibilities and in this case the leaving variable is $x2$. Note that the leaving variable corresponds to the minimum ratio of (3.17) and (3.18).

3. If equations (3.17) - (3.18) and (3.21) - (3.22) were not present, then the leaving variable would be $x4$ in keeping with the second objetive of removing as many infeasibilities as possible. Note that the leaving variable corresponds to the maximum ratio of (3.19) - (3.20).

The above example suggests the need for classifying rows/ratios into three types:

Type 1 : If both pivot and rhs of a row are greater than 0 then we call such a row type 1. Its ratio is called a type 1 ratio. The type 1 ratio should be a min ratio just like the normal RSM ratio test.

Type 2 : If both pivot and rhs for a row are less than 0 and the basic variable for the row is not an equality logical then we call such a row type 2. Its ratio is called a type 2 ratio. The ratio for a type 2 row should be a max ratio due to the second objective of removing as many infeasibilities as possible.

Type 3 : If both pivot and rhs of a row are less than 0 and the basic variable for the row is an equality logical then we call such a row type 3. The ratio for such a row is called a type 3 ratio. The ratio for a type 3 row should be a min ratio due to the first objective.

Type ignored : If the pivot and rhs of a row are of opposite sign then the basic variable for such a row would not be considered to leave the basis since no bound is provided by such a row.

To summarize:

RHS	Pivot	Variable Type	Action	Type	Ratio Name
+	+	Any type	Choose Min	1	DRMIN
+	-	Any type	Ignore	--	--
-	+	Any type	Ignore	--	--
-	-	Except equality logical	Choose Max	2	DRMAX
-	-	Equality logical	Choose Min	3	DRART

Thus, at the end of one pass through all the rows there will in general be three ratios DRMIN, DRMAX and DRART correponding to the three types. The type of the pivot row is then decided by :

1. If there are no rows of type 1 or type 3 then the pivot row is of type 2 and is decided by DRMAX due to the second objective.

2. If there are rows of type 1 or type 3 then the pivot row is decided by min{DRMIN, DRART}.

3. If there are no rows of types 1, 2 or 3 then all rows have rhs and pivots with opposite signs and the problem is unbounded.

As mentioned earlier, CHUZR is called to find the max pivot row (when JENTRY = 2) from among those pivots which provide the same best change in the objective value. To find the max pivot row given the inputs mentioned above a scan is done through all the rows of the pivot column having the same type as indicated by NTYPE. The value of the max pivot (MAXPIV) is initialized to zero and then a scan is performed as explained below. Consider the ith row of type NTYPE. If the ratio of the row does not exceed DE, the max pivot row is set to i provided the value of its pivot element exceeds MAXPIV. MAXPIV is then set to the value of this pivot. Otherwise, the next row is considered. For example, assume that in (3.21) the coefficient of x1 is -6 and the rhs is -24. If the pivot column is the column corresponding to x1 then the ratio is of type 3. This would have been decided during the first call to CHUZR from NORMAL. When CHUZR is called again with JENTRY set to 2, the pivot row is given by (3.21), since the coefficient of x1 in (3.21) is higher than the coefficient of x1 in (3.22).

3.5.6. Steps

If not called with a flag (JENTRY ≠ 2) then

1. Compute the type of the pivot row for the given candidate column.

4. Compute the change in the objective value.

Else

2. Find the best eligible pivot.

3. Check for negative right-hand side (rhs) elements and make appropriate

changes.

    End if

3.5.7. Detailed Comments

```
C************************************************************************
C     Step 1 :                                                    CHUZR
C
C     YA(*, *)  : Buffer containing a pool of candidate columns.
C     KVEC      : Column of YA that has a candidate column currently being considered.
C     JENTRY    : Flag. '2' indicates that the maximum pivot is to be found.
C                  Otherwise, the type of the pivot row is to be found.
C     X(*)      : Array containing the solution vector.
C     NROW      : Number of rows.
C
C     1. If JENTRY is not 2 then for each row i, do:
C               -  If there is a zero in the pivot position of the i^th row then go to the next row as
C                  such a row does not provide a bound.  [line 7 - 9].
C                  Else do
C               -  Set IVEC to the index of the basic column for the i^th row. [line 10]
C               -  If row i is non-binding go to the next row since the basic variable of a non-
C                  binding row is never a candidate to leave the basis. [line 11 - 12]
C                  If binding then
C
C                  Case 1 :  (Pivot and rhs are > 0)
C                            YA(I, KVEC) > 0 implies pivot > 0. [line 13]
C                            X(I) > 0 implies rhs > 0. [line16]
C                            Since this is the case of a type 1 ratio, compute the ratio DTHETA
C                            and update the type 1 min ratio DRMIN. [lines 17 - 20]
C
C                  Case 2 :  (Pivot and rhs < 0 and for inequality logical )
C                            YA(I, KVEC) < 0 implies pivot < 0. [line 14]
C                            IVEC < NROW implies equality logical and
C                            ISTYPE(IVEC) ≠ -1 implies inequality logical. [line 21 - 22]
C                            X(I) < 0 implies rhs is < 0. [line 23]
C                            Compute the ratio DTHETA and since this is the case of a type 2
C                            ratio, update the type 2 max ratio DRMAX. [line 24 - 28]
C
```

```
C            Case 3 :  (Pivot and rhs < 0 for equality logical)
C                      ISTYPE(IVEC) = -1 implies equality logical.  [line 22]
C                      Compute the ratio and since this is the case of a type 3 ratio and
C                      update the type 3 min ratio DRART. [line 30 - 33]
C          End If
C   End For
C
C   2. The type of the pivot row is decided as follows.
C
C            Case 1 :  (No type 1 or type 2 or type 3 rows)
C                      This is indicated by DRMIN, DRMAX, DRART remaining at their
C                      initialized values. Note that even though these values were initialized
C                       to 1.0D30 and -1.0D30 they can only be tested for 1.0D29 and
C                      -1.0D29 due to the precision limit on the computer.  [lines 34, 42]
C                      Therefore, all rows have pivots and rhs with opposite signs.  Go
C                      to step 2 to set the status to "unbounded".  [line 42]
C
C            Case 2 :  (No type 1 or type 3 rows)
C                      Line 34 tests for this case.  The ratio type is 2 and DRMAX gives
C                      the row corresponding to this type.  [lines 43 - 46]
C
C            Case 3 :  (Type 1 or type 3 rows exist)
C                      Line 34 tests for this case.  Pivot row type is decided according to
C                      min{DRMIN, DRART}.  [lines 35 - 40]
C
C          Go to step 2 to find the change in the objective value.  [line 41]
C
C*********************************************************************
C                                                            CHUZR
        IF(JENTRY .EQ. 2) GO TO 1400                           1
        DRMIN = 1.D30                                          2
        IROWP = 0                                              3
        DRMAX = -1.D30                                         4
        DRART = 1.D30                                          5
        DO 1000 I = 1, NROW                                    6
        IF(DABS(YA(I, KVEC)) .GT. ZTOLZE) GO TO 50             7
        YA(I, KVEC) = 0.0                                      8
```

```
C                                                              CHUZR
        GO TO 1000                                               9
  50    IVEC = JH(I)                                             10
        IF ( IVEC .GT. NROW) GO TO 100                           11
        IF(ISTYPE(IVEC) .EQ. 0) GO TO 1000                       12
 100    IF(YA(I, KVEC) .GT. ZTOLPV) GO TO 150                    13
        IF(YA(I, KVEC) .LT.-ZTOLPV) GO TO 500                    14
        GO TO 1000                                               15
 150    IF(X(I) .LT. -ZTOLRJ) GO TO 1000                         16
        DTHETA = (X(I) + ZTOLRJ) / YA(I, KVEC)                   17
        IF(DTHETA .GE. DRMIN) GO TO 1000                         18
        DRMIN = DTHETA                                           19
        GO TO 1000                                               20
 500    IF(IVEC .GT. NROW) GO TO 710                             21
        IF(ISTYPE(IVEC) .EQ. -1) GO TO 800                       22
 710    IF(X(I) .GE. 0) GO TO 1000                               23
        DTHETA = X(I) / YA(I, KVEC)                              24
        IF (DTHETA .LE. DRMAX) GO TO 1000                        25
        DRMAX = DTHETA                                           26
        IROWP2 = I                                               27
        GO TO 1000                                               28
 800    IF(X(I) .GT. ZTOLRJ) GO TO 1000                          29
        DTHETA = (X(I) - ZTOLRJ) / YA(I, KVEC)                   30
        IF(DTHETA .GE. DRART) GO TO 1000                         31
        DRART = DTHETA                                           32
1000    CONTINUE                                                 33
C
        IF(DRMIN .GT. 1.D29 .AND. DRART .GT. 1.D29) GO TO 1300   34
        IF(DRART .LT. DRMIN) GO TO 1200                          35
        NTYPE = 1                                                36
        DE = DRMIN                                               37
        GO TO 9900                                               38
1200    DE = DRART                                               39
        NTYPE = 3                                                40
        GO TO 9900                                               41
1300    IF(DRMAX .LE. -1.D29) GO TO 9000                         42
        DE = DRMAX                                               43
```

```
C                                                                  CHUZR
        NTYPE = 2                                                     44
        IROWP = IROWP2                                                45
        GO TO 9900                                                    46
C*********************************************************************
C        Step 2 :
C
C        YA(*, KVEC)    : Contains the pivot column.
C        X(*)           : Array containing the solution vector.
C        DE             : Starts with the allowable increase in the entering variable after first
C                          pass of all rows.  As each row is considered during the second call
C                          to CHUZR with JENTRY = 2, DE gets  updated to indicate the
C                          allowable increase in the entering variable at  that point.
C        DRO            : Initialized to allowable increase in the entering variable.
C        DTHETA         : Ratio for the row being considered.
C        NTYPE          : Pivot row's type set during the previous call with JENTRY ≠ 2.
C
C        For each row i row do :
C
C           Case 1  : (Pivot < 0)
C           -  YA(I, KVEC) < 0 implies this case.  [line 53]
C           -  If NTYPE ≠ 3 then NTYPE should be 1 and and the pivot of the min ratio's
C              row was > 0.  Therefore, the i^th row could not be the one that set the pivot
C              row's type. Go to the next row.
C              Else NTYPE is 3:  [line 62]
C              - If the basic variable is not an equality logical or if the rhs is > 0 then the i^th
C                row cannot be of type 3.  Therefore go to the next  row.
C              Else proceed.  [lines 63 - 65]
C              - Find the ratio DTHETA for the i^th row.  If DTHETA > DRO  then the i^th row
C                 does not give the min ratio.  Go to the next  row. Else update the value of the
C                 maximum pivot (MAXPIV) and IROWP to i to indicate that so far the i^th row
C                 has yielded the maximum pivot.  [lines 66 - 71]
C
C           Case 2  :  (Pivot > 0)
C              This is the complement of case 1.  [lines 54 - 60, 70 - 71]
C
C        End For
```

```
C                                                                          47
C***********************************************************************
C                                                                    CHUZR
1400      MAXPIV = 0.0                                                    47
          DRO = DE                                                       48
          DO 2000 I = 1, NROW                                            49
          IVEC = JH(I)                                                   50
          IF(IVEC .GT. NROW) GO TO 1450                                  51
          IF(ISTYPE(IVEC) .EQ. 0) GO TO 2000                             52
1450      IF(YA(I, KVEC) .LE. ZTOLPV) GO TO 1500                         53
          IF(NTYPE .NE. 1) GO TO 2000                                    54
          IF(X(I) .LT. -ZTOLRJ) GO TO 2000                              55
          DTHETA = X(I) / YA(I, KVEC)                                    56
          IF(DTHETA .GT. DRO) GO TO 2000                                 57
          IF(YA(I, KVEC) .LT. MAXPIV) GO TO 2000                         58
          MAXPIV = YA(I, KVEC)                                           59
          GO TO 1600                                                     60
1500      IF(YA(I, KVEC) .GE. -ZTOLPV) GO TO 2000                        61
          IF(NTYPE .NE. 3) GO TO 2000                                    62
          IF(IVEC .GT. NROW) GO TO 2000                                  63
          IF(ISTYPE(IVEC) .NE. -1) GO TO 2000                            64
          IF(X(I) .GT. ZTOLRJ) GO TO 2000                                65
          DTHETA = X(I) / YA(I, KVEC)                                    66
          IF(DTHETA .GT. DRO) GO TO 2000                                 67
          IF(-YA(I, KVEC) .LT. MAXPIV) GO TO 2000                        68
          MAXPIV = - YA(I, KVEC)                                         69
1600      DE = DTHETA                                                    70
          IROWP = I                                                      71
2000      CONTINUE                                                       72
C***********************************************************************
C         Step 3 :
C
C         If pivot row is of type 3 then the rhs of the pivot row must be ≤ 0. Likewise if the
C         pivot row is of type 1 then the rhs of the pivot row must be ≥ 0. [lines 73 - 81]
C
C         Step 4 :
C
```

C	Set MSTAT = 'QU' if the problem is unbounded. [line 82]	
C	Compute the change DY in the objective value by multiplying the reduced cost	
C	DCMIN(KVEC) of the entering column with the allowable increase DE in the	
C	entering variable. [line 84]	
C		

```
************************************************************************
C                                                                  CHUZR
           IF(NTYPE .EQ. 3) GO TO 3000                                73
           IF(X(IROWP) .GE. 0) GO TO 10000                            74
           DE = 0                                                     75
           X(IROWP) = 0                                               76
           GO TO 10000                                                77
 3000      IF(X(IROWP) .LT. 0) GO TO 10000                            78
           DE = 0                                                     79
           X(IROWP) = 0                                               80
           GO TO 10000                                                81
 9000      MSTAT = QU                                                 82
           DE = 1.E10                                                 83
 9900      DY = DCMIN(KVEC)*DE                                        84
 10000     RETURN                                                     85
           END                                                        86
```

3.6. Subroutine FORMC

3.6.1. Major Task

Form the basic cost vector for each simplex iteration.

3.6.2. Inputs

The solution vector in the array X(*).

3.6.3. Outputs

1. Forms the basic cost vector into the appropriate column of the buffer named YA(*, *).

2. Also sets variables to denote the feasibility of the current solution.

3.6.4. Called by

FTRAN, NORMAL.

3.6.5. Description

FORMC sets up the cost vector in a buffer named YA(*, *). To do so we first determine which column of YA(*, *) is to store the cost vector and then the contents of the cost vector.

The column of the buffer YA to store the cost vector depends on a variable called KFASE. KFASE is normally set to '2' in which case the cost vector is set up in YA(*, 6). When the solution of a subproblem becomes infeasible, KFASE is set to '1' and the cost vector is set up in YA(*, 5).

The contents of the buffer depends on the status of the problem and is given as follows.

Case 1 : (Feasible solution and executing step 1 of D - W Phase 1) Reference to §2.6.1 shows that the cost vector for a feasible solution in a subproblem should be $[1, -\pi_0^k, 0]$. Since π_0^k is a $\mathbf{0}$ vector to start with, the cost vector is $[1, \mathbf{0}, \mathbf{0}]$ for $k = 0$.

Case 2 : (Feasible solution and executing D-W Phase 3) In this case the objective function is fixed (cf. § 2.5.3). Therefore, the cost vector for this case is $[1, \mathbf{0}, \mathbf{0}]$ which is the cost vector for an LP in RSM Phase 2.

Case 3 : (Feasible solution and executing either step 2 of D-W Phase 1 or D-W Phase 2)
  Case 3.1 : (Solving master problem) In this case the objective function is fixed. Therefore, the cost vector is $[1, \mathbf{0}, \mathbf{0}]$.
  Case 3.2 : (Solving subproblem) From § 2.6.1 it can be seen that the cost vector is $[1, -\pi_0^k, 0]$. Note that the 1 and $-\pi_0^k$ are stored in the sixth column of YA(*, *) by the subroutine MASTER before FORMC is called.

Case 4 : (Infeasible solution in any phase) Cost vector corresponds to infeasibilities as discussed in § 2.2.3.

3.6.6. Steps

  1. Set the appropriate buffer vector and determine if step 2 should be performed. If not go to 3.
  2. Checks the solution vector for infeasibilities and produces the cost vector accordingly.
  3. Find the cost vector for feasible cases.

3.6.7. Detailed Comments

```
C***********************************************************************
C     Step 1:                                                    FORMC
C
C     NROWO  : Number of coupling rows.
C     MAST   : Stage of the D-W algorithm that is currently being executed.
C                '0' indicates step 1 of D-W Phase 1.
C                '1' indicates step 2 of D-W Phase 1 or D-W Phase 2.
C                '2' indicates D-W Phase 3.
C     KFASE  : Normally set to 2. Set to 1 if solving a subproblem in D-W Phase 2
C                and the subproblem's solution is not feasible.
C     LSUB   : Index of the problem being treated.  '0' indicates the master problem.
C     IFFEZ  : Status of the current solution.  Feasible if '1' and not feasible if '0'.
C     YA(*, *) : Buffer in which the cost vector is set up.
C
C              -  Initialize IFFEZ to 1 indicating feasible status.  [line 1]
C                 Initialize KVEC to  6 indicating that the cost vector should be formed in
C                 YA(*, 6).  [line 2]
C
C              - Case 1 : (MAST = 1)
C                 Case 1.1 : (Solving subproblem) In this casethe value of KFASE can be used
C                  to indicate the status (feasible or infeasible) of the problem's solution.
C
C                  - If KFASE = 1 then set KVEC to 5 to set up the cost vector in YA(*, 5).
C                   [line 9]
C                  -Set YA(*, 5) to the zero vector.  [lines 11 - 13]
C                  - Proceed to step 2 to check if all infeasibilities have been removed and if so
C                   set the objective row's position in YA(*, *) to 1.
C                   Reference to subroutine NORMAL will show that in this case FORMC would
C                   be called again with KFASE set to 2.
C                   On the other hand if all  infeasibilities have not been removed then  step 2 is
C                   used to set the cost vector according to the infeasibilities.
C                  - If KFASE = 2 then reference to subroutine MASTER will show that the
C                   values of [1, -π0k] are already stored in the first NROWO positions of
C                   YA(*, 6). Since the required cost C vector is [1, -π0k, 0], set each of the
C                   elements from YA(NROWO+1, 6) to YA(NROW, 6) to zero. [lines 5 - 9]
C
C                 Case 1.2 : (Solving master problem)
```

```
C               - Set YA(*, 6) to zero . [lines 11 - 13]
C               - Go to step 2 to check for any infeasibilities. If none,  step 3 will set [1, 0, 0]
C                  as the cost vector.
C           Case 2 :  (MAST = 0 or 2)
C               Same as case 1.2.
C*************************************************************************
 C                                                              FORMC
        IFFEZ = 1                                                 1
        KVEC = 6                                                  2
        IF(MAST .NE. 1) GO TO 50                                  3
        IF(LSUB .LT. 1 .OR. KFASE .NE. 2) GO TO 25                4
        MM = NROWO + 1                                            5
        DO 2500 I = NROWO+1                                       6
  2500  YA(I, KVEC) = 0.                                          7
        GO TO 9000                                                8
   25   IF(KFASE .EQ. 1) KVEC = 5                                 9
   50   MSTAT = QF                                               10
        DO 100 I = 1, NROW                                       11
        YA(I, KVEC) = 0.                                         12
  100   CONTINUE                                                 13
        SUM = 0.                                                 14


C*************************************************************************
C       Step 2 :
C
C       For each row i do
C               - Set ICOL to the index of the basic column in the ith row.  [line 16]
C           Case 1 :  (non-binding row)
C               - ISTYPE(*) = 0 implies non-binding row.  [line 18]
C               - Ignore and go to the next row.
C
C           Case 2 :  (structural or inequality logical)
C               - ICOL > NROW implies structural.  [line 17]
C               - ICOL ≤ NROW and ISTYPE(*) = +1 implies inequality logical.
C                [lines 17 - 18]
C               - If X(i) < 0 then store a +1.0 in YA(i, KVEC). [lines 24 - 25]
C               - Update SUM, the sum of infeasibilities.  [line 26]
```

65

```
C           - Set IFFEZ to 0 indicating that the current solution is not feasible.  [line 27]
C
C           Case 3 :  (equality logical)
C           - ICOL ≤ NROW and ISTYPE(*) = -1 implies equality logical. [lines 17 - 18]
C           - If X(i) < 0 then YA(i, *) = +1
C             If X(i) > 0 then YA(i, *) = -1.  This is done because an artificial has to be
C             driven to zero.  In either case update SUM and set IFFEZ to 0 indicating
C             that the current solution is infeasible.  [lines 19 - 23, 27 - 28]
C
C           End For
C*********************************************************************************
C                                                                         FORMC
            DO 1000 I = 1, NROW                                             15
            ICOL = JH(I)                                                    16
            IF (ICOL .GT. NROW) GO TO 500                                   17
            IF (ISTYPE(ICOL)) 200, 1000, 500                               18
  200       IF (DABS(X(I)) .LE. ZTOLRJ) GO TO 1000                         19
            IF(X(I) .LT. 0.) YA(I, KVEC) = +1.                             20
            IF(X(I) .GT. 0.) YA(I, KVEC) = -1.                             21
            SUM = SUM + DABS(X(I))                                          22
            GO TO 510                                                       23
  500       IF(X(I) .GT. -ZTOLRJ) GO TO 1000                              24
            YA(I, KVEC) = +1.                                               25
            SUM = SUM - X(I)                                                26
  510       IFFEZ = 0                                                       27
            MSTAT = QI                                                      28
 1000       CONTINUE                                                        29
C*********************************************************************************
C     Step 3 :
C
C     Based on the results of step 2 there are two cases :
C
C     Case 1 :  (Feasible solution)
C     - IFFEZ = 1 implies feasible solution.  [line 31]
C     - Set the element in YA corresponding to the objective row to 1.0.  [line 32]
C     Case 2 :  (Not feasible)
C     - IFFEZ = 0 implies infeasible solution.  [line 31]
```

```
C          - RETURN.
C**************************************************************************
C                                                                    FORMC
          SUMINF = SUM                                                  30
          IF (IFFEZ .LE. 0) GO TO 9000                                  31
          YA(IOBJ, KVEC) = 1. 0                                         32
9000      RETURN                                                        33
```

3.7. Subroutine FTRAN

3.7.1. Major Task

Performs the column update $d_j = E_k(... (E_2 (E_1 a_j)))$ where a_j is the original column and d_j its update. Also gives the reduced cost if called with a suitable flag.

3.7.2. Inputs

    1. The eta file consisting of etas in the array A(*).

    2. The column to be updated in YA(*, KVEC).

    3. The type of FTRAN indicated by the parameter IPAR.

3.7.3. Outputs

    1. The updated column in YA(*, KVEC).

    2. Also computes the reduced cost if the parameter IPAR is 3.

3.7.4. Called by

NORMAL, INVERT

3.7.5. Calling

Calls FORMC to form c_B if IPAR is 3 (flag to compute reduced cost).

3.7.6. Description

FTRAN performs two functions. The first is to update a given column a_j to d_j by solving the system $B d_j = a_j$. The second is to find the reduced cost if FTRAN is called with a flag.

Updating a column is done by transforming it with a set of elementary eta matrices as shown by (3.23)

$$d_j = E_k...E_1 a_j \qquad\qquad (3.23)$$

where

$$B = F_1...F_k \qquad\qquad (3.24)$$

and

$$F_i^{-1} = E_i \qquad\qquad (3.25)$$

Before this transformation can be started, one should know the etas necessary for the required update and this depends on the control parameter IPAR. Three cases arise. The first is a full update in which the transformation is done with all the etas generated so far. The second is a partial update used in factorizing the "bump" in INVERT. The third is a single step update used in multiple pricing in which certain columns have to be updated by the eta created in the previous simplex pivot. Let E_g be the gth eta and let the pth column of E_g be the non-trivial column.

$$E_g = [u_1,..., n_p,..., u_m] \qquad\qquad (3.26)$$

where

$$n_{ip} = \begin{array}{ll} - e_{ip}/e_{pp} & i \neq p \\ 1/e_{pp} & i = p \end{array} \qquad\qquad (3.27)$$

and u_i is the ith unit vector.

Without loss of generality consider the transformation by E_g. Note that at this point the original column a_j has already been transformed to d'_j (say) by eta matrices $E_1,..., E_{g-1}$. Therefore, the gth stage of the trannsformation is :

$$E_g d'_{ij} = \begin{array}{ll} d'_{ij} - e_{ip} d'_{pj}/e_{pp} & i \neq p \\ d'_{pj}/e_{pp} & i = p \end{array} \qquad\qquad (3.28)$$

Since the pivot element of the eta is stored first (cf. §3.21), the row index of the first element gives the pivot row p of E_g. Therefore, the pth component (= d'_{pj}/e_{pp}) of the update can be computed easily. As can be seen from (3.28), computing each of the other elements of the update requires only one multiplication and one addition.

As stated earlier, the reduced cost has also to be computed if FTRAN is called with the parameter IPAR set to '3'. To compute the reduced cost (cf. § 2.2.3)

$$\pi a_j = c_B B^{-1} a_j = c_B d_j \qquad (3.29)$$

one needs to multiply the appropriate cost vector (c_B) to the updated column d_j just obtained. This is done as follows:

Case 1 : (Solution feasible)

 Case 1.1 : (Solving master problem) In this case the objective function is fixed and therefore the cost vector (cf. § 2.2.3) is [1, **0**, **0**]. Hence, the reduced cost is the first element of d_j.

 Case 1.2 : (Solving subproblem in step 1 of D-W Phase 1 or Phase 3) Same as case 1.1.

 Case 1.3 : (Solving subproblem in D-W Phase 2) The cost vector in this case is [1, $-\pi_0^k$, **0**] and reference to subroutine MASTER shows that this is stored in the sixth column of the array YA(*, *). The reduced cost is then computed by the inner product given by (3.29).

Case 2 : (Previous solution infeasible) In multiple pricing (cf. § 2.2.4) FORMC ia not called after the solution gets updated due to a basis change. Since the simplex method maintains feasibility, if the previous solution was feasible, then the solution after an update is also feasible. But if the previous solution was infeasible then FTRAN has to call FORMC to ascertain the status of the current solution. Therefore two cases result from this call to FORMC.

 Case 2.1 : (Current solution still infeasible) The reduced cost is the inner product of the cost vector for infeasibilities just set up by FORMC and the updated column d_j.

 Case 2.2 : (Current solution feasible) FORMC is called again to set up the appropriate cost vector for feasible solutions and then case 1 is executed.

3.7.7. Steps
 1. Find the etas necessary for the update.
 2. Perform the update.
 3. Find the reduced cost if flagged.

3.7.8. Detailed comments

```
C**********************************************************************
C    Step 1 :                                                   FTRAN
C
```

```
C     IPAR     : Determines the type of FTRAN to be performed. If '1' then a full update.
C                If '2' then partial update for factorization of the "bump" in INVERT. If '3'
C                then single pivot update using the latest eta.
C     NFE      : First eta for the update. If IPAR = 1 then NFE = 1. If IPAR = 2 then NFE is
C                index of first eta in the "bump". If IPAR = 3 then NFE is last eta generated.
C     NLE      : Last eta of the update. This is always the last eta in the eta file.
C
C*****************************************************************************
C                                                                     FTRAN
      CALL TICTAC(JTIM)                                                  1
      GO TO (100, 110, 120), IPAR                                        2
  100 NFE = 1                                                            3
      NLE = NETA                                                         4
      GO TO 200                                                         5
  110 NFE = NLETA + 1                                                    6
      NLE = NETA                                                         7
      GO TO 200                                                         8
  120 NFE = NETA                                                         9
      NLE = NETA                                                        10


C*****************************************************************************
C     Step 2 :
C
C     LL        : First non-zero element of the non trivial column of the eta being used.
C     KK        : Last non-zero element of the non trivial column of the eta being used.
C     IA(*)     : Row index of the (*)th element of the data matrix A.
C     YA(*, *)  : Buffer, one of whose columns requires the update.
C     KVEC      : The column of YA that is to be updated.
C     DY        : The pivot element of the updated column.
C     A(*)      : Contains the eta file.
C
C     1. If there are no etas then RETURN. Else proceed. [line 11]
C     2. For each eta Ei do :
C
C          -  Set LL and KK to the first and last position of the eta np which represents Ei
C             [line 13-14]
C          -  Since the pivot element of the eta is stored first the row index of the first
```

```
C              non-zero in n_p is p. Set IPIV to p to indicate pivot row.
C              If the pivot is close to zero then take up the next eta.  [line 15 - 16]
C          -   DY is the pivot element of the updated column d_j since A(LL) gives the pivot
C              element 1/e_pp of the eta and YA(IPIV, KVEC) gives the pivot element a_pj
C              of the column a_j.  [line 17]
C          -   Set pivot element of the updated column.  [line 18]
C          -   From (3.28) it is clear that if an eta element e_ip is zero then there would be no
C              change in the corresponding updated element of d_j.  So for each non-zero eta
C              element j of E_i, find the row index IR of the j^th element and update the IR^th
C              position of YA(*, *) as given by (3.28).
C
C          End For
C
C************************************************************************
C                                                             FTRAN
200        IF (NFE .GT. NLE) GO TO 9000                        11
           DO 1000 IK = NFE, NLE                               12
           LL = LE(IK)                                         13
           KK = LE(IK + 1) - 1                                 14
           IPIV = IA(LL)                                       15
           IF(DABS(YA(IPIV, KVEC)) .LE.1.D-15) GO TO 1000      16
           DY = YA(IPIV, KVEC)*A(LL)                           17
           YA(IPIV, KVEC) = DY                                 18
           IF (KK .LE. LL) GO TO 1000                          19
           LL = LL + 1                                         20
           DO 500 J = LL, KK                                   21
           IR = IA(J)                                          22
           YA(IR, KVEC) = YA(IR, KVEC) - A(J)*DY               23
500        CONTINUE                                            24
1000       CONTINUE                                            25
C************************************************************************
C          Step 3 :
C
C          MSTAT    : Denotes solution status. 'QF' for feasible, 'QI' for infeasible.
C          NROW     : Number of rows.
C          NROWO    : Number of coupling rows.
```

```
C    YA(*, *)    : Contains the updated column whose reduced cost is to be computed.
C    KVEC        : The column of YA whose reduced cost is to be found.
C    KFASE       : Normally set to '2' in for the cost vector to be set up in YA(*, 6).
C                  When a subproblem is not feasible in D-W Phase 2 then KFASE is set
C                  to '1' and the cost vector is set up in YA(*, 5).
C    MAST        : Denotes the stage of the D-W algorithm being executed.
C                  If '0' then executing step 1 of D-W Phase 1.
C                  If '1' then executing step 2 of D-W Phase 1 or D-W Phase 2.
C                  If '2' then executing D-W Phase 3.
C    LSUB        : Index of the problem being solved.
C                  '0' indicates MASTER problem.
C    IPAR        : If '3' then FTRAN is flagged to find the reduced cost.
C
C    1. If not flagged then RETURN. Else proceed. [line 26]
C    2. The cases cited below are the same as those given in "description".
C      Case 1 : (Solution feasible)
C
C      MSTAT = 'QF' implies feasible solution. [line 27]
C        Case 1.1 : (Solving master problem) Reduced cost is the first element.
C      [line 47]
C        Case 1.2 : (Solving subproblem in step 1 of Phase 1 or Phase 3 of D-W)
C      Same as case 1.1
C        Case 1.3 : (Solving subproblem in D-W Phase 2) Reduced cost is the
C      inner product of the subproblem's D-W Phase 2 cost vector in YA(I, KK) and the
C      updated column in YA(I, KVEC). [line 43 - 46]
C      MM in line 41 is set to NROWO since the subproblem's Phase 2 cost vector is
C      zero beyond the NROWO$^{th}$ position. [line 44]
C      KK is set to 6 since the Phase 2 cost vector is in YA(*, 6)
C
C      Case 2 : (Previous solution not feasible) MSTAT ≠ QF implies this case. [line 27]
C      Call FORMC to ascertain current solution's status. [line 28]
C      After calling FORMC there are two cases. [line 29]
C
C        Case 2.1 : (Solution still infeasible) Reduced cost is now the inner product of
C      the cost vector for infeasibilities in YA(I, KK) and the updated column
C      YA(I, KVEC). [lines 44-46] KK is set according to KFASE. [lines 31 - 32]
C      MM of line 44 is set to the number of rows NROW, since any of the NROW
```

```
C        positions of the solution vector X(*) can be infeasible.  [line 30]
C
C           Case 2.2  :  (Solution feasible)  If KFASE is 2 then the cost vector is already set up
C        in YA(*, 6).  Therefore, find reduced cost as in case 1. [line 34]
C        If KFASE is 1 then in the previous iteration the subproblem's solution was infeasible
C        and the cost would have been formed in YA(*, 5).  Hence, set KFASE to 2 and call
C        FORMC to set up the cost vector in YA(*, 6).
C        Proceed to Case 1 to compute the reduced cost.  [lines 34 - 38]
C
C*************************************************************************
C                                                                      FTRAN
         IF(IPAR .LT. 3) GO TO 9000                                      26
         IF(MSTAT .EQ. QF) GO TO 1200                                    27
         CALL FORMC                                                      28
         IF(MSTAT .EQ. QF) GO TO 1100                                    29
         MM = NROW                                                       30
         KK = 6                                                          31
         IF(KFASE .EQ. 1) KK=5                                           32
         GO TO 1400                                                      33
1100     IF(KFASE .NE. 1) GO TO 1200                                     34
         KFASE = 2                                                       35
         CALL FORMC                                                      36
         DOB = 0                                                         37
         DO 1150 I = 1, NROWO                                            38
1150     DOB = DOB - X(I)*YA(I, 6)                                       39
1200     IF(LSUB .EQ. 0 .OR. MAST .NE. 1) GO TO 2000                     40
         MM = NROWO                                                      41
         KK = 6                                                          42
1400     DCMIN(KVEC) = 0                                                 43
         DO 1500 I = 1, MM                                               44
1500     DCMIN(KVEC) = DCMIN(KVEC) + YA(I, KK)*YA(I, KVEC)               45
         GO TO 9000                                                      46
2000     DCMIN(KVEC) = YA(IOBJ, KVEC)                                    47
900      CALL TICTAC(ITINV)                                              48
         JTINV = JTINV + ITINV - JTIM                                    49
         RETURN                                                          50
         END                                                            51
```

3.8. Subroutine INDATA

3.8.1. Major Task
Executes Step 1 of D-W Phase 1.

3.8.2. Called By
Main Program

3.8.3. Calling
1. INPUT to read the problem.
2. NORMAL to solve subproblems.
3. VECTOR to write to and read from the direct access device (DAD).
4. PACK to pack the proposals.

3.8.4. Description
INDATA executes step 1 of D-W Phase 1. This is done in three stages. The first is the set-up stage; the second is the proposal generation stage; and the third stage sets up the master problem for step 2 of D-W Phase 1.

To set up the problem, INDATA has to first read the control parameters. Since DECOMP is design to solve large problems, it is assumed that data for the complete problem cannot be accommodated in core at the same time. Therefore, a direct access device (DAD) consisting of a fixed number of records of fixed length is defined to store the data relevant to each problem. The organization of DAD is shown in Figure 3.1. The number of records required to write the problem's names (row and column names) and data is determined. Using this, the elements of a two dimensional pointer array NOLA(*, *) (cf. § 5.5.2) are computed. This concludes the set-up stage.

The second stage consists of execution of the following instructions LMAXP1 (= number of subproblems plus one) times.

- A problem (master or subproblem) is read from the input file.

Case 1 : (Do not generate proposals)

- This case is met if the problem read is the master problem or if the user desires that no proposal be generated in step 1 of D-W Phase 1.

- Since the master problem is not solved at this point, the problem's data is written to DAD.

- Each of the coupling rows is marked as non-binding.

- The number of rows (NROW) is initialized to the number of coupling rows (NROWO). This is done since the subproblem structure (cf. § 2.6.1) includes the coupling rows as non-binding rows.

    Case 2 : (Generate proposals)

    - NORMAL is called to solve the subproblem. If it is infeasible then the whole problem is infeasible. Otherwise, a column with the proposal (modulo a minus sign), is generated and kept in the first NROWO elements of the solution array X(*). This is copied to the proposal buffer Z(*, *).

    - The problem's data is written to DAD so as to free the space in core for reading and solving the next subproblem.

    - The number of rows (NROW) is set to the number of coupling rows (NROWO) for the same reason as cited in case 1 above.

    - The names in the master problem are read from DAD. This is done to check the existence of row names given for data in $A1_r$ (cf § 2.5.3).

    The last stage is to read the master problem's data from DAD. PACK is called to incorporate the proposals into the data array of the master problem so that step 2 of D-W Phase 1 can be executed.

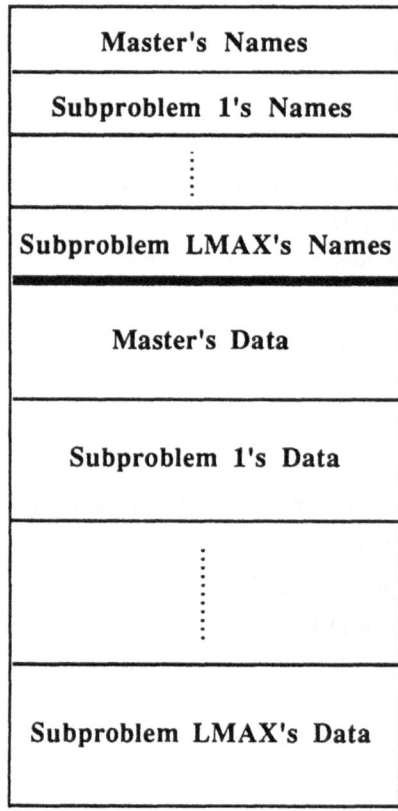

Figure 3.1 Organization of the direct access file

3.8.5. Steps

1. Set up.
2. Solve all problems and generate proposals if necessary.
3. Pack proposals.

3.8.6. Detailed Comments

```
C*************************************************************************
C    Step 1 :                                                     INDATA
C
C    A(*)    :  Array containing the matrix elements and the eta file of a problem.
C    LPOINR :  Indicates where in A(*) the eta file of a problem begins.
C    NEMAX  :  Maximum number of non-zero elements allowed in the array A(*) of the
C                 problem. Set to 10,000.
C    NRMAX  :  Maximum number of rows in a problem.
C    NLAMAX :  Maximum number of columns in a problem.
C    NTMAX  :  Maximum number of eta vectors in the eta file of a problem.
C    LTRAC  :  Record length in bytes of a record in the direct access file.
C    MAXTRK :  Maximum number of records defined in the direct access file.
C    NTRAC2 :  Number of records in the direct access file required to store the names
C                 (row and column) of any subproblem.
C    NTRAC1 :  Number of records in the direct access file required to store all data relevant
C                 (other than names) to any subproblem.
C    ITRK   :  Total number of records in the direct access file required to store all relevant
C                 data for the entire problem (master and subproblems).
C    Z(*, *) :  Proposal buffer.
C    NPROS  :  Column dimension of Z(*, *).
C    MAXCA  :  Maximum number of proposals sent to the master by a particular problem in
C                 a given cycle.
C    MAXNCA :  Control parameter which gives the maximum number of proposals that can
C                 be sent to the master in a cycle. Default is 100 and maximum is 150.
C    KMULT  :  Maximum number of columns in candidate pool in multiple pricing.
C
C    1. Define the unit 9 direct access file.  [line 6]
C    2. Initialize problem parameters.  [lines 7 - 37]
C    3. If parameters are supplied by the user then read them.  [lines 40 - 42]
C    4. If the parameters are outside the allowable range then set them to the appropriate
C       values. [lines 43 - 45]
```

```
C    5. Compute NTRAC1 and NTRAC2.  [lines 47 - 48]
C    6. Based on the above, compute for each subproblem (including master), the starting
C       record number in the direct access file for its data.  Store this information in
C       NOLA(*, *). Also compute ITRK which indicates the total number of records required
C       for the entire problem.  [lines 49 - 60]
C    7. If ITRK exceeds the maximum number of tracks defined for the direct access file
C       (MAXTRK) then STOP.  Otherwise proceed to step 2. [lines 68 - 71]
C
C**********************************************************************
C                                                             INDATA
         NAMELIST/PARAM/, LPOINR, LMAX, IOBJ, USERHS, KINP,         1
       1 KSTR, MAXNCA, MAXCA, NCANFR, INVFRQ, INVFMA, NPH,
       2 KMULT, ITOL, ICPI, IRST, ICLAST, ISAVE, ISOL, RPERC, RSTOP
         NAMELIST /TOL/ ZTOLZE, ZTOLPV, ZTCOST, ZTOLRJ, ZTCAND      2
         DIMENSION NB(3612)                                         3
         EQUIVALENCE (ICNAM(1), A(5001)), (JH(1), NB(1))            4
         DATA MAXTRK /101/                                          5
         DEFINE FILE 9 ( 78, 3257, U, KD)                           6
         REWIND 8                                                   7
         QPROP = QBL                                                8
         IPROS = 0                                                  9
         MAST = 0                                                  10
         DO 90 I = 1, 6                                            11
   90    IPROP(I) = 0                                              12
         LPOINR = 5000                                             13
         IOBJ = 1                                                  14
         USERHS = 1                                                15
         KINP = 5                                                  16
         KSTR = 2                                                  17
         MAXNCA = 100                                              18
         MAXCA = 10                                                19
         NCANFR = 15                                               20
         INVFRQ = 30                                               21
         INVFMA = 20                                               22
         NPH = 1                                                   23
         KMULT = 3                                                 24
         ITOL = 0                                                  25
```

```
C                                                              INDATA
        ICPI = 0                                                  26
        IRST = 0                                                  27
        ICLAST = 999                                              28
        ISAVE = 1                                                 29
        ISOL = 0                                                  30
        RPERC = 40.                                               31
        RSTOP = 0.01                                              32
        ZTOLZE = 1.0E-07                                          33
        ZTOLPV = 1.0E-06                                          34
        ZTCOST = 1.0E-04                                          35
        ZTOLRJ = 1.0E-05                                          36
        ZTCAND = 1.0E-02                                          37
C
        CALL TICTAC(ITIMIN)                                       38
        NROW = 0                                                  39
        OPEN (2, FILE = 'DECOMP.DAT', STATUS = 'OLD')             40
        READ(2, PARAM)                                            41
        IF (ITOL .NE. 0) READ(5, TOL)                             42
        IF (KSTR .EQ. 5 .AND. MAXCA .GT. NPROS) MAXCA = NPROS     43
        IF(MAXNCA .GT. 150) MAXNCA = 150                          44
        IF (KMULT .LE. 0 .OR. KMULT .GT. 5) KMULT = 1             45
        LMAXP1 = LMAX + 1                                         46
        NTRAC1 = (10*NEMAX + 4*(NRMAX + NLAMAX)                   47
       1 +2*NTMAX+LTRAC+19)/LTRAC
        NTRAC2 = (8*NLAMAX + LTRAC + 7)/LTRAC                     48
        NOLA(3, LMAXP1) = 1 + NTRAC2*LMAXP1                       49
        NOLA(4, LMAXP1) = 1                                       50
        IF (LMAX .EQ. 0) GO TO 105                                51
        DO 100 I = 1, LMAX                                        52
        NOLA(3, I) = NOLA(3, LMAXP1) + NTRAC1*I                   53
100     NOLA(4, I) = NOLA(4, LMAXP1) + NTRAC2*I                   54
        ITRK = NOLA(3, LMAX) + NTRAC1 - 1                         55
        GO TO 110                                                 56
105     ITRK = NTRAC2 + NTRAC1                                    57
110     NOLA(1, LMAXP1) = ITRK + 1                                58
        ITRK = ITRK + (2*NLAMAX + 303)/LTRAC + 1                  59
```

```
C                                                                    INDATA
        WRITE(6, 6010) LMAX, LPOINR, IOBJ, USERHS, KINP,                60
      1 KSTR, MAXNCA, MAXCA, NCANFR, INVFRQ, INVFMA, NPH,
      2 KMULT, ITOL, ICPI, IRST, ICLAST, ISAV, ISOL, NTRAC1,
      3 NTRAC2, ITRK, RPERC, RSTOP
 6010   FORMAT('1PROGRAM PARAMETERS' // '0LMAX', T8, I4 / '             61
      1 LPOINR', T8, I4 / ' IOBJ', T8, I4 / ' USERHS', T8, I4 / ' KINP', T8, I4 /
      2 'KSTR', T8, I4 / ' MAXNCA', T8, I4/' MAXCA', T8, I4 / ' NCANFR', T8, I4 /
      3 ' INVFRQ', T8, I4 / ' INVFMA', T8, I4 / ' NPH', T8, I4 / ' KMULT', T8, I4 /
      4 ' ITOL', T8, I4 / ' ICPI', T8, I4/' IRST', T8, I4 / ' ICLAST', T8, I4 /
      5 ' ISAVE', T8, I4 / ' ISOL', T8, I4 / ' NTRAC1', T8, I4 /
      6 ' NTRAC2', T8, I4 / ' ITRK', T8, I4 / ' RPERC   ', G10.3, ' PERCENT' /
      7 ' RSTOP   ', G10.3, ' PERCENT')
        IF (ZTCAND .LE. ZTCOST) ZTCAND = ZTCOST                         62
        WRITE (6, 6015) ZTOLZE, ZTOLPV, ZTCOST, ZTOLRJ, ZTCAND          63
 6015   FORMAT (' ZTOLZE ',G10.3 / ' ZTOLPV ', G10.3 /                  64
      1 ' ZTCOST ', G10.3 / ' ZTOLRJ ', G10.3 / ' ZTCAND ', G10.3)
        RCTEST = -1.0E+05*ZTCOST                                        65
        RPERC = 0.01*RPERC + 1.                                         66
        RSTOP = 0.01*RSTOP                                              67
        IF (ITRK .LE. MAXTRK) GO TO 112                                 68
        WRITE (6, 6020) ITRK                                            69
 6020   FORMAT('0INSUFFICIENT STORAGE SPACE ON UNIT 9.', I5,            70
      1 ' TRACKS NEEDED')
        STOP                                                            71
 112    IF (IRST .NE. 0) GO TO 1500                                     72


C********************************************************************
C       Step 2 :
C
C       LMAX        : Number of subproblems.
C       LMAXP1      : Number of subproblems plus one.
C       LSUB        : Index of the subproblem being considered.
C       Z(*, *)     : Proposal buffer.
C       NROW        : Number of rows in the subproblem being considered.
C       NROWO       : Number of coupling rows.
C       NPROS       : Column dimension of Z(*, *).
```

```
C     IPROS        : Column index in Z(*, *) where the next proposal is to be written.
C     NPH          : Control parameter only for step 1 of D-W Phase 1. '1' indicates
C                     each subproblem is to be solved to optimality to generate proposals.
C     X(*)         : Array containing the solution of the problem being considered.
C     MSTAT        : Status of the problem being considered.  QN indicates that the
C                     problem is infeasible.
C
C     A problem is read from the input file.  There are now two cases :
C
C          Case 1  : (Do not generate proposals) If the problem read is the master problem
C     or if NPH is not one this case is met.  The data relevant to the problem is written to
C     the direct access  device. [lines 82 - 85, 101 - 103]
C          If the problem read is the master problem then mark the coupling rows as
C     "non-binding" and incorporate the coefficient of the logicals (ones) in the array A(*).
C     [lines 106 - 108]
C          Set NROW to NROWO. Since the subproblem data contains the coupling rows
C     as non-binding rows, this is necessary to keep a correct tally for the number of rows
C     in a subproblem.  [lines 110]
C          Read the names (row and column) of the master problem.  This is done to
C     check that the coupling rows in the subproblem's data have been declared.
C     [lines 112 - 113]
C          Case 2  : (Generate proposals)  NORMAL is called to solve the subproblem.
C     If the subproblem is infeasible then the whole problem is infeasible.  In this case
C     STOP.  [lines 87 - 91]
C          Otherwise a proposal would have been generated.  In this case increment
C     IPROS, the column pointer in Z(*, *) at which the proposal is to be written.  The first
C     NROWO rows of X(*) contain the proposal, modulo a minus sign. Hence copy
C     -X(*) to Z(*, IPROS).  [lines 92 - 98]
C          Write the data relevant to the problem to the unit 9 file.  [lines 101 - 103]
C     Set NROW to NROWO.  Also, read the names of the master problem.  These are
C     done for the same reason as explained in case 1.
C
C*****************************************************************************
C                                                                  INDATA
      DO 1000 I = 1, LMAXP1                                        73
      IF (I .LE. LMAX) MSTATU(I) = QF                              74
      LSUB = I - 1                                                 75
```

```
C                                                               INDATA
        CALL INPUT                                                  76
        ITCNT = 0                                                   77
        NT = 0                                                      78
        NX = 0                                                      79
        ITSINV = 0                                                  80
        KK = LSUB                                                   81
        IF(LSUB .EQ. 0)KK = LMAXP1                                  82
        KD = NOLA(4, KK)                                            83
        WRITE(9'KD) NTEMP(1), NTEMP(2), ICNAM                       84
        LCORE = LSUB                                                85
        IF(LSUB .EQ. 0 .OR. NPH .NE. 1) GO TO 500                   86
        CALL NORMAL                                                 87
        IF(MSTAT .NE. QN) GO TO 300                                 88
        WRITE(6, 200) LSUB, SUMINF                                  89
200     FORMAT(//, ' SUBPROBLEM ', I4, ' HAS NO SOLUTION.          90
       1 SUM OF INFEASIBILITIES = ', F16.8)
        STOP                                                        91
300     IPROS = IPROS + 1                                           92
        IF (IPROS .LE. NPROS) GO TO 330                             93
        WRITE (8) Z                                                 94
        IPROS = 1                                                   95
330     DO 350 J=1, NROWO                                           96
350     Z(J, IPROS) = -X(J)                                         97
        Z(NROWO + 1, IPROS) = 0.0                                   98
500     KD = NOLA(3, KK)                                            99
C
        IF(NELEM. GT. (NEMAX-NRMAX) .OR.                           100
       1 ITSINV.GT.(4*INVFRQ) / 5) NETA = 0
        NOLA(2, KK) = LE(NETA + 1) - 1                             101
        KIKA = 1                                                   102
        CALL VECTOR(KIKA, A, IA, NB, NOLA(2, KK))                  103
600     IF(LSUB .GT. 0) GO TO 800                                  104
        IF(LMAX .EQ. 0) GO TO 1000                                 105
        DO 750 J = 1, NROWO                                        106
        ISTYPE(J) = 0                                              107
750     A(J) = 1.                                                  108
```

```
C                                                                    INDATA
800       IF(LSUB .EQ. LMAX) GO TO 1000                              109
          NROW = NROWO                                               110
          IF (LSUB .EQ. 0 .OR. NPH .NE. 1) GO TO 1000                111
          KD = NOLA(4, KK)                                           112
          READ(9'KD) NTEMP(1), NTEMP(2), ICNAM                       113
1000      CONTINUE                                                   114
```

```
C*********************************************************************
C       Step 3 :
C
C       A(*)          : Array which has the data and eta elements of the problem.
C       NOLA(*, *)    : NOLA(3, *) indicates the record number on the direct access file
C                       where tha data and the eta elements of the *th problem are stored.
C                       NOLA(2, *) indicates the position in the array A(*) where the eta
C                       elements of the *th problem end.
C       KIKA          : "1" indicates that the data is to be written on the direct access file and
C                       "2" indicates that the data is to be read from the direct access file.
C       LPOINR        : Indicates the position in the array A(*) where the eta elements begin.
C       LE(1)         : Indicates where in the array A(*) the eta file of the problem begins.
C       NPH           : Control parameter used only in step 1 of D-W Phase 1.
C                       If not "1" it indicates that no proposals are to be generated in step 1
C                       of D-W Phase 1. Otherwise proposals are to be generated only at
C                       optimality.
C       NCAND         : Number of proposals generated in the current cycle.
C       LMAX          : Number of subproblems.
C       NROWO         : Number of coupling rows.
C       NCOLO         : Number of coupling columns.
C       JH(*)         : Indicates the column that is basic in the *th row.
C       KINBAS(*)     : Indicates if *th column is basic and if so the row in which it is basic.
C
C       1. Obtain the record number of the direct access file where the problem's data and eta
C          are stored.  [line 115]
C       2. Set KIKA to 2 indicating that the data is to be read from the direct access device.
C          [line 118]
C       3. Call VECTOR to read the data from the direct access device.  [line 119]
C       4  If NPH is not one then no proposals would be generated and therefore the master
```

82

```
C       problem is ready to be solved. In this case prepare to exit subroutine.  Otherwise
C       proceed as follows.  [line 121]
C   5. Since problems are solved sequentially and since only one proposal is generated at
C       optimality, the jth proposal is from the jth subproblem.  Hence assign J to
C       NAME(J). Also set NCAND to LMAX since one proposal has been generated per
C       subproblem.  Call PACK to incorporate the proposals in the array A(*).
C       [lines 122 - 125]
C   6. The convexity constraint corresponding to each subproblem has only one column
C       and therefore it has to be basic. Set JH(*) and KINBAS(*) to indicate this.
C       [lines 129 - 134]
C
C*****************************************************************************
C                                                                   INDATA
        KD = NOLA(3, LMAXP1)                                          115
        LSUB = 0                                                      116
        LCORE = 0                                                     117
        KIKA = 2                                                      118
        CALL VECTOR(KIKA, A, IA, NB, NOLA(2, LMAXP1))                 119
        LE(1) = LPOINR                                                120
        IF(NPH .NE. 1 .OR. LMAX .EQ. 0) GO TO 1500                    121
        DO 1100 J = 1, LMAX                                           122
 1100   NAME(J) = J                                                   123
        NCAND = LMAX                                                  124
        CALL PACK                                                     125
        NCAND = 0                                                     126
        MM = NCOLO                                                    127
        NN = NROWO                                                    128
        DO 1200 J = 1, LMAX                                           129
        NN = NN + 1                                                   130
        MM = MM + 1                                                   131
        JH(NN) = MM                                                   132
        KINBAS(NN) = 0                                                133
        KINBAS(MM) = NN                                               134
 1200   CONTINUE                                                      135
 1500   CALL TICTAC(JTIMIN)                                           136
        TIMER = (JTIMIN - ITIMIN) / 1000.                             137
        WRITE(6, 2000)TIMER                                           138
```

2000	FORMAT(// '0TOTAL INPUT TIME = ', F8.2, ' SECONDS', /, 1H1)	139
	IF(LMAX .GE. 1) MAST = 1	140
	RETURN	141
	END	142

3.9. Subroutine INPUT

3.9.1. Major Task

Read an LP in MPS format.

3.9.2. Input

A file called DECOMP.DAT that contains the problem parameters and the problem data. An example of the input file of a typical problem is given in the user's guide.

3.9.3. Called By

INDATA.

3.9.4. Detailed Comments :

```
C*******************************************************************
C                                                              INPUT
C    LSUB     : Index of the problem that is to be read in.
C    NROW     : Number of rows in the problem being read.
C    NROWO  : Number of coupling rows.
C    A(*)      : Data array.
C    KINBAS(*): Indicates if the *th column is basic and if so the row in which it is basic.
C    JH(*)     : Indicates the column that is basic in the *th row.
C    LMAX     : Number of subproblems.
C    NCOL     : Number of columns (including rhs) in the problem being read.
C    NELEM   : Number of non-zeros in the problem being read.
C    USERHS : Control parameter. DECOMP allows for the possibility  of reading in many
C                     right-hand sides and the USERHSth rhs is used.
C    IVIN      : Number of structural columns in the "columns" section of the problem.
C
C    1. While not end of file: read a card.
C    2. If the card just read isan "END" go to step 3.
```

84

```
C                                                                              INPUT
C        Otherwise there are several cases:
C             Case 1 :  (NAMES card)
C        - Record the name in NTEMP.
C        - Write the card.
C             Case 2 :  (ROWS card)
C        - Set L = 1 to flag the start of the "ROWS" section.
C        - Write the card.
C             Case 3 :  (COLUMNS card)
C        - Set L = 2 to flag the start of the COLUMNS section.
C        - If reading the master problem then
C             - The rows listed in the "rows" section of the master problem are the coupling
C               rows.  Hence set NROWO to NROW.  [line 50]
C             - An index is maintained to indicate the number of binding coupling rows.  If
C               there are none then the entire problem breaks up into LMAX independent
C               problems. Hence, solving all subproblems to optimality would provide the
C               optimal solution. This can be achieved by setting KSTR to one. [lines 53 - 58]
C             - Incorporate the convexity rows into A(*).  Also, mark the type of the convexity
C               rows as binding equality rows.  [lines 59, 62 - 67]
C             - Increment NROW by LMAX since the master problem has LMAX convexity
C               conatriants.  [lines 59, 60]
C             - Name the ith convexity row QQi. [lines 68 - 69]
C           End if
C        - At this point the problem has only logicals. Therefore, it has as many columns and
C          as many non-zero elemens as there are rows. Hence, set both NCOL and NELEM
C          to NROW.  [lines 71 - 72]
C             Case 4 :  (RHS card)
C        - Set L = 4 to signify the start of the RHS section.
C        - Set IVIN to NCOL. Hence IVIN indicates the current number of columns in
C          the columns section of the problem.  [line 134]
C        - Write the card.
C             Case 5  :  (L = 1 implying ROWS section card)
C        - Ignore coupling rows included in the ROWS section of a subproblem.
C          [lines 25 - 27]
C        - Increment row count (NROW).  [line 28]
C        - Record row name and row type.  [lines 30 - 31]
C        - Incorporate logical coefficient in A(*) according to row type.
```

```
C          A(*) is -1 for "≥" rows and 1 otherwise.  [lines 32 - 49]
C        - Record KINBAS(*) and JH(*).
C            Case 6 :  (L = '2' or '4' implying COLUMNS or RHS section card)
C        - If the column name on the current card is not the same as that on the previous card
C          then it implies the start of a new column. [line 80]
C          In this case :
C            - Test for split vector ((i.e) test if part of this column was read in before). If so
C               print a message.  [lines 79 - 84]
C            - Increment the number of columns (NCOL).  [line 85]
C            - Record column name.  [lines 86 - 87]
C            - Set LA(NCOL) to NELEM.  Hence, LA(NCOL) indicates where in A(*)
C               the NCOL^th column begins. [line 92]
C          End If
C        - If the row name on the current card matches one of the names given in the "ROWS"
C          section of the master problem then the non-zero element on this card is part of the
C          data.  Hence incorporate the data into the array A(*) and increment NELEM.
C          [lines 92 - 112]
C
C  3.  Once the "END" card is read then
C        - Set LA(NCOL+1) to NELEM+1. This signifies the end of data in all NCOL
C          columns.  [line 134]
C        - Since the eta file for every problem except the master starts at the end of the
C          problem's data,  LE(1) is set to NELEM+1.  [line 135]
C        - IVIN indicates the number of columns after reading the data in the columns section.
C          DECOMP allows for the possibility of choosing among various right hand sides.
C          Hence, by setting RHSCOL to IVIN+USERHS the user can specify which rhs to
C          use. [line 137]
C        - If master problem then initialize the subproblem origin of the columns.
C        - Print problem statistics.
C
C******************************************************************************
C                                                                        INPUT
       EQUIVALENCE (ICNAM(1), A(5001))                                      1
       CALL TICTAC(JTIM)                                                    2
       WRITE(6, 6000)                                                       3
6000   FORMAT(1H1)                                                          4
   5   READ(KINP, 700) K1, K2, K3, K4, (NAME(I), I = 1, 4),                 5
```

```
C                                                                        INPUT
      1 ATEMP1, NAME(5), NAME(6), ATEMP2
700   FORMAT(4A1, 2A4, 2X, 2A4, 2X, G12.6, 3X, 2A4, 2X, G12.6)            6
      IF(K1 .EQ. QBL) GO TO 50                                           7
      IF(K1 .EQ. QN) GO TO 100                                          8
      IF(K1 .EQ. QR .AND. K2 .EQ. QO) L = 1                             9
      IF(K1 .EQ. QR .AND. K2 .EQ. QO) GO TO 150                         10
      IF(K1 .EQ. QC) L = 2                                              11
      IF(K1 .EQ. QC) GO TO 300                                          12
      IF(K1 .EQ. QB .AND. K2 .EQ. QA) L = 3                             13
      IF(K1 .EQ. QB .AND. K2 .EQ. QA) GO TO 150                         14
      IF(K1 .EQ. QR .AND. K2 .EQ. QH) L = 4                             15
      IF(K1 .EQ. QR .AND. K2 .EQ. QH) GO TO 500                         16
      IF(K1 .EQ. QE) GO TO 600                                          17
50    GO TO(210, 320, 410, 320), L                                     18
100   NTEMP(1) = NAME(3)                                               19
      NTEMP(2) = NAME(4)                                               20
150   WRITE(6, 710) K1, K2, K3, K4, NAME(1), NAME(2), NAME(3), NAME(4)  21
710   FORMAT(3X, 4A1, 2A4, 2X, 2A4)                                    22
      GO TO 5                                                          23
C
210   IF (LSUB .LE. 0) GO TO 215                                       24
C
      DO 212 I = 1, NROWO                                              25
      IF (ICNAM(I, 1) .NE. NAME(1) .OR.                                26
      1 ICNAM(I, 2) .NE. NAME(2)) GO TO 212
      GO TO 5                                                          27
212   CONTINUE                                                         28
215   NROW = NROW + 1                                                  29
      ICNAM(NROW, 1) = NAME(1)                                         30
      ICNAM(NROW, 2) = NAME(2)                                         31
C
C     TEST ROW TYPE
C
      IF(K2 .EQ. QL .OR. K3 .EQ. QL) GO TO 220                         32
      IF(K2 .EQ. QE .OR. K3 .EQ. QE) GO TO 230                         33
      IF(K2 .EQ. QG .OR. K3 .EQ. QG) GO TO 240                         34
```

```
C                                                              INPUT
        IF(K2 .EQ. QN .OR. K3 .EQ. QN) GO TO 250                35
        GO TO 230                                                36
220     ISTYPE(NROW) = 1                                         37
        GO TO 255                                                38
230     ISTYPE(NROW) = -1                                        39
        GO TO 255                                                40
240     ISTYPE(NROW) = 1                                         41
        A(NROW) = -1.                                             42
        GO TO 260                                                43
250     ISTYPE(NROW) = 0                                         44
255     A(NROW) = 1.                                             45
260     IA(NROW) = NROW                                           46
        LA(NROW) = NROW                                           47
        JH(NROW) = NROW                                           48
        KINBAS(NROW) = NROW                                       49
        GO TO 5                                                  50
C
C       MATRIX ELEMENTS
C
300     IF(LSUB .GT. 0 .OR. LMAX .EQ. 0) GO TO 310               51
        NROWO = NROW                                              52
        J = 0                                                    53
        DO 305 I = 1, NROW                                        54
        IF(ISTYPE(I) .EQ. 0) GO TO 305                            55
        J = J + 1                                                56
305     CONTINUE                                                 57
        IF(J .EQ. 0) KSTR = 1                                     58
        DO 308 I = 1, LMAX                                        59
        NROW = NROW + 1                                           60
        ISTYPE(NROW) = - 1                                        61
        A(NROW) = 1.                                             62
        IA(NROW) = NROW                                           63
        LA(NROW) = NROW                                           64
        ICOLS(NROW) = 0                                           65
        JH(NROW) = NROW                                           66
        KINBAS(NROW) = NROW                                       67
```

```
C                                                                        INPUT
        ICNAM(NROW, 1) = QQ                                                68
        ICNAM(NROW, 2) = I                                                 69
308     CONTINUE                                                          70
310     NCOL = NROW                                                       71
        NELEM = NROW                                                      72
        ICS1 = QQ                                                         73
        ICS2 = QQ                                                         74
        GO TO 150                                                        75
320     J = 3                                                            76
        K = 4                                                            77
        IF (NAME(1) .EQ.ICS1 .AND. NAME(2) .EQ.ICS2) GO TO 328            78
C
C       TEST FOR SPLIT VECTOR
C
        DO 325 I = 1, NCOL                                                79
        IF (NAME(1) .NE. ICNAM(I, 1)) GO TO 325                           80
        IF (NAME(2) .NE. ICNAM(I, 2)) GO TO 325                           81
        WRITE(6, 8250) NAME(1), NAME(2)                                   82
8250    FORMAT(14H0SPLIT VECTOR , 2A4)                                    83
325     CONTINUE                                                         84
        NCOL = NCOL + 1                                                   85
        ICS1 = NAME(1)                                                    86
        ICS2 = NAME(2)                                                    87
        ICNAM(NCOL, 1) = ICS1                                             88
        ICNAM(NCOL, 2) = ICS2                                             89
        LA(NCOL) = NELEM + 1                                              90
        KINBAS(NCOL) = 0                                                  91
C
C       TEST FOR ROW MATCH
C
328     IF (DABS(ATEMP1) .GT. ZTOLZE) GO TO 330                           92
        J = 5                                                            93
        K = 6                                                            94
        ATEMP1 = ATEMP2                                                   95
330     DO 340 I = 1, NROW                                                96
        IF (DABS(ATEMP1) .LE. ZTOLZE) GO TO 335                           97
```

```
C                                                                    INPUT
        IF(NAME(J) .NE. ICNAM(I, 1) .OR. NAME(K) .NE. ICNAM(I, 2))      98
      1 GO TO 340
        IF (L .EQ. 4 .AND. LSUB .GT. 0 .AND. I .LE. NROWO) GO TO 335    99
        NELEM = NELEM + 1                                              100
        IA(NELEM) = I                                                  101
        A(NELEM) = ATEMP1                                              102
  335   IF (K .GT. 5) GO TO 5                                          103
        IF(DABS(ATEMP2) .LE. ZTOLZE) GO TO 5                          104
        J = 5                                                          105
        K = 6                                                          106
        ATEMP1 = ATEMP2                                                107
        GO TO 330                                                      108
  340   CONTINUE                                                       109
        WRITE(6, 8300) NAME(J), NAME(K)                               110
 8300   FORMAT(18H0NO MATCH FOR ROW , 2A4)                            111
        STOP                                                           112
C
C       BASIS CARDS
C
  410   DO 420 I = 1, NCOL                                             113
        IF(NAME(1) .NE. ICNAM(I, 1) .OR.                               114
      1 NAME(2) .NE. ICNAM(I, 2)) GO TO 420
        IBVEC = I                                                      115
        GO TO 425                                                      116
  420   CONTINUE                                                       117
        WRITE(6, 8400) NAME(1), NAME(2)                               118
 8400   FORMAT(21H0NO MATCH FOR VECTOR , 2A4)                         119
        GO TO 5                                                        120
  425   DO 430 I = 1, NROW                                             121
        IF(NAME(3) .NE. ICNAM(I, 1) .OR.                               122
      1 NAME(4) .NE. ICNAM(I, 2)) GO TO 430
        IBROW = I                                                      123
        GO TO 440                                                      124
  430   CONTINUE                                                       125
        WRITE(6, 8300)NAME(3), NAME(4)                                126
        GO TO 5                                                        127
```

```
C                                                             INPUT
440     JH(IBROW) = IBVEC                                       128
        KINBAS(IBROW) = 0                                       129
        KINBAS(IBVEC) = IBROW                                   130
        GO TO 5                                                 131
C
C       RHS
C
500     IVIN = NCOL                                             132
        GO TO 150                                               133
C
C       END OF INPUT
C
600     LA(NCOL+1) = NELEM + 1                                  134
        LE(1) = NELEM + 1                                       135
        NETA = 0                                                136
        RHSCOL = IVIN + USERHS                                  137
        NUMRHS = NCOL - IVIN                                    138
        IF(LSUB .GT. 0) GO TO 615                               139
        DO 605 I = 1, NCOL                                      140
605     ICOLS(I) = 0                                            141
        MM = IVIN + 1                                           142
        DO 610 I=MM, NCOL                                       143
610     KINBAS(I) = NROW + 1                                    144
        NCOLO = NCOL                                            145
        NSCOL = IVIN - NROW                                     146
        GO TO 620                                               147
615     NCOL = IVIN                                             148
        NSCOL = NCOL - NROW                                     149
620     NELEM = LA(IVIN+1) - 1 - NROW                           150
        RELEM = NELEM                                           151
        RDENS = 100.                                            152
        IF(NSCOL .EQ. 0)GO TO 625                               153
        RDENS = RELEM / (NROW*NSCOL)*100.                       154
625     WRITE(6, 710) K1, K2, K3, K4, NAME(1), NAME(2)          155
        WRITE(6, 8500) NROW, NSCOL, NELEM, RDENS                156
8500    FORMAT(19H0PROBLEM STATISTICS / 1H , I5, 5H ROWS / 1H , I5,   157
```

```
C                                                            INPUT
          1 19H STRUCTURAL COLUMNS / 1H I5, 18H NON-ZERO
          2 ELEMENTS / 211H DENSITY = , F6.2, ' PERCENT')
            CALL TICTAC(ITIM)                               158
            TIMER = (ITIM - JTIM) / 1000.                   159
            WRITE(6, 8900) NTEMP(1), NTEMP(2), TIMER        160
8900        FORMAT('0INPUT TIME SUBPROBLEM ', 2A4,          161
          1 ' = ', F6.2, 'SECONDS')
            RETURN                                          162
            END                                             163
```

3.10.   Subroutine INVERT

3.10.1. Major Task

Computes the L-U factorization of the basis.

3.10.2. Inputs

1. The array JH(i) which specifies the column that is basic in the ith row.
2. The array KINBAS(i) which tells if the ith column is basic and if so the row in which it is basic.
3. The basic columns which reside in the array A(*).

3.10.3. Outputs

1. Eta vectors which define the L and U factors of the basis **B**.
2. The updated solution vector X(*).

3.10.4. Description

INVERT applies a series of symbolic permutations to the basis. Roughly speaking, this rearranges **B** so that it is "almost" lower triangular and that only a small sub-matrix of **B** would have to be factorized. INVERT consists of five steps :

1. A pivot sequence is decided for all row and column singletons. A row (or column) singleton is a row (or column) having just one non-zero element after the removal of rows and columns corresponding to all previously assigned pivots.

2. The resulting basis matrix is then permuted to an easily factorizable form. Note that no physical permutation of the data is actually performed. A pivot sequence specifying which row and column to use in each pivot will be determined. This pivot sequence is equivalent to a

permutation of the matrix and then pivoting down the diagonal. For ease of exposition, we shall use "permutation" and "assignment of pivot sequence" interchangeably.

3. At the end of steps 1 and 2, it will be seen later that only a sub-matrix **B2** of **B** (called the bump) would have to be factorized. Step 3 consists of computing the L-U factors of the bump **B2**.

4. Form etas.

5. Compute the solution.

The matrix is conceptually partitioned into four. The first partition called part 1 contains row singletons. The second and third partitions called part 3 and part 4 respectively contain column singletons. The unnamed partition is the bump **B2**. Step 1 is known as pre-assigned pivot procedure and is based on Hellerman and Rarick's method [1971].

Step 1 Pre-assigned Pivot sequence :

Step1 is done in three stages.

1.1 Put logicals in part 4 : A scan of all the rows is performed as follows. Consider the ith row. The basic variable of each row is tested to see if it is a logical. If so, rows and columns are permuted so that the logical appears on the diagonal of part 4. Otherwise, the column belongs tentatively to part 1 and hence the number of columns in part 1 is incremented. At the end of step 1.1 a general basis as shown in Figure 3.2 has the structure as shown in Figure 3.3. Note that **permutations are never carried out physically**. Instead, the effect of a permutation is achieved by changing the order of pivoting. This is done through two arrays MREG(i) and VREG(i) which indicate that the ith pivot will be in row MREG(i) and column VREG(i). For example, consider the matrix in Figure 3.4. A logical would be detected while considering the third row. MREG(5) is set to 4 and VREG(5) is also set to 4 indicating that the 5th pivot will be in row 4 and column 4. This is shown in Figure 3.5.

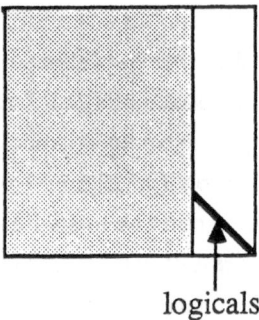

logicals

Figure 3.2 The basis as a sparse matrix   Figure 3.3 The basis after step 1.1

	1	2	3	4	5
1	x	x	x		
2	x				
3		x			x
4	x		x	x	x
5		x	x		

Figure 3.4 An example basis

	1	2	3	5	4
1	x	x	x		
2	x				
3		x		x	
5		x	x		
4	x		x	x	x

Part 4

Figure 3.5 Example basis after step 1.1

1.2 Establish row counts : First, rows and columns corresponding to all singleton elements of part 4 are marked as "unavailable" (assigned for pivot). Then all columns of the tentative part 1 are scanned as follows. Without loss of generality, consider the jth part 1 column. The row index (say i) of a non-zero in the column is obtained. If the ith row was not assigned for pivot in part 4, then the number of non-zeros in that row is updated. On the other hand, if the ith row was assigned for pivot then it is ignored. The number of such unassigned non-zeros in the column is updated. This is repeated for all non-zeros in that column. If the column has only one non-zero in unassigned rows, then a column singleton has been detected and in this case rows and column are permuted so that the singleton element appears on the diagonal of part 3. At the end of this pass through the columns, the non-zero count of each row is obtained. At the end of step 1.2, the matrix would have the structure as shown in Figure 3.6. For the example given in step 1.1, column 5 is a column singleton. Therefore, MREG(4) is set to 3 and VREG(4) is set to 5 to give the effect of a permutation. This is shown in Figure 3.7.

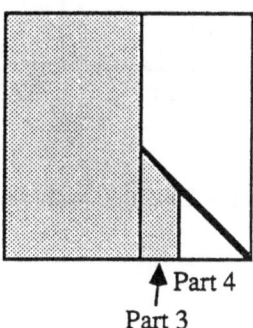

Part 4

Part 3

Figure 3.6 The basis after step 1.2

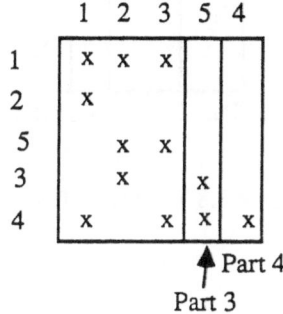

	1	2	3	5	4
1	x	x	x		
2	x				
5		x	x		
3		x		x	
4	x		x	x	x

Part 4

Part 3

Figure 3.7 Example basis after step 1.2

1.3 Put row singletons in part 1 and new column singletons in part 3 : This step is a multi-pass scan through part 1 columns until all row and column singletons have been removed. Every time a pass is completed, another pass is undertaken provided at least one row or column singleton was detected during the previous pass. In each pass all part 1 columns are scanned to put row singletons in part 1 and column singletons in part 3 as explained below. Consider the jth part 1 column during the pth pass. The row index (say i) of a non-zero in the column is obtained. If row i has been assigned for pivot it is ignored. If the row was not assigned and is a row singleton (recall step 1.2 gives the non-zero count of all part 1 rows), then

(a) rows and columns are permuted so that the singleton element appears on the diagonal of part 1;

(b) the row and column corresponding to the singleton element are marked as "unavailable" (assigned for pivot) and then the next ($(j+1)^{st}$) part 1 column is taken up for consideration.

This procedure is repeated for all non-zeros in the jth column . If no row singletons were detected and if the number of non-zeros in unassigned rows of the column is one, then a column singleton has been detected. In this case, rows and columns are permuted so that the singleton element appears on the diagonal of part 3. Again, the row and column corresponding to the singleton element are marked as "unavailable". At the end of this step the matrix would have the structure as shown in Figure 3.8. For the example, the structure looks as shown in Figure 3.9.

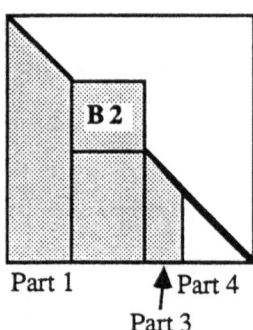

Figure 3.8 The basis after step 1.3

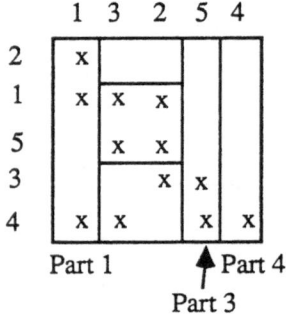

Figure 3.9 Example basis after step 1.3

Step 2 Bringing the basis to an easily factorizable form :

This is illustrated in Figures 3.10 - 3.12. Knowing the L-U factors of the bump **B2** (=**L2'U2'** say), the L-U factors of the matrix given in Figure 3.12 is shown in Figure 3.13. At the end of this step, the eta file contains already the columns corresponding to m1 as L etas and the columns corresponding to m3 as U etas.

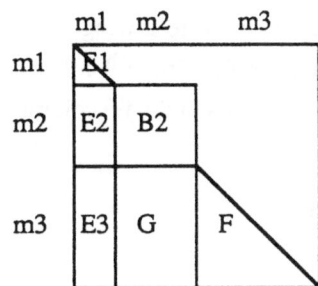

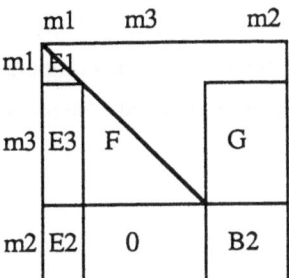

Figure 3.10 The basis after step 1.   Figure 3.11 First rearrangement in step 2

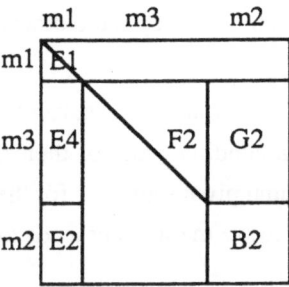

Figure 3.12 The basis after step 2

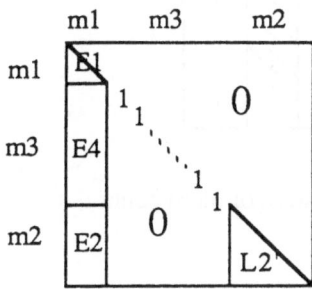
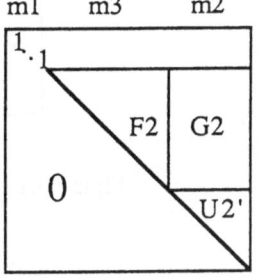

Figure 3.13 L-U factors of the basis.

Step 3 Factorizing the bump **B2** :

First a pivot sequence has to be established to minimize the amount of fill-ins. Then the factors are obtained using Gaussian elimination and the corresponding etas are written to the array A.

3.1 Pivot Sequence for the bump : This is based on the following heuristics. For each column, a "merit count" is computed. The merit count $m(j)$, of the jth column, is the sum of the number of unassigned non-zeros in rows having non-zeros in the jth column. Thus

$$m(j) = \sum_i (r_i f_{ij})$$

where

    r_i is the number of unassigned non-zeros in row i; and

    f_{ij} is 1 if row i has a non-zero in column j and 0 otherwise.

For example, the merit count of column 4 is seven for the matrix shown in Figure 3.14. Next, the columns are sorted in ascending order of their merit counts using shell sort. This order will then be used as the column pivot sequence for the bump. The heuristic argument for this choice is that a column with a lower merit count tends to creat less fill-in when pivoted.

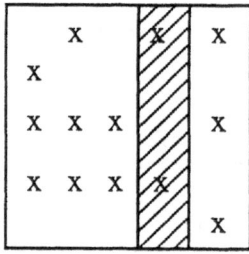

Figure 3.14 Illustration of merit count

3.2 L-U factors of the bump : This is done by Gaussian elimination. The jth step of Gaussian elimination which consists of pre-multiplying **B2** by the elementary matrix $\mathbf{L2''_j}$ is as shown in Figure 3.15. Thus, at the end of m2 (since dimension of **B2** is m2 by m2) Gaussian elimination steps we have :

$$\mathbf{L2''_1}\ \mathbf{L2''_2}...\ \mathbf{L2''_{m2}}\ \mathbf{B2}\ =\ \mathbf{U2'} \tag{3.30}$$

$$\mathbf{B2} = [\mathbf{L2'}_{m2} \; ... \; \mathbf{L2'}_1] \, [\, \mathbf{U2'}_1 \; ... \; \mathbf{U2'}_{m2} \,] \qquad\qquad (3.31)$$

where $\mathbf{U2'}_j$ $(j = 1,..., m2)$ is the elementary matrix whose j^{th} column is the jth column of $\mathbf{U2'}$ and $\mathbf{L2'}_j = \mathbf{L2''}_j^{-1}$ $(j = 1,..., m2)$.

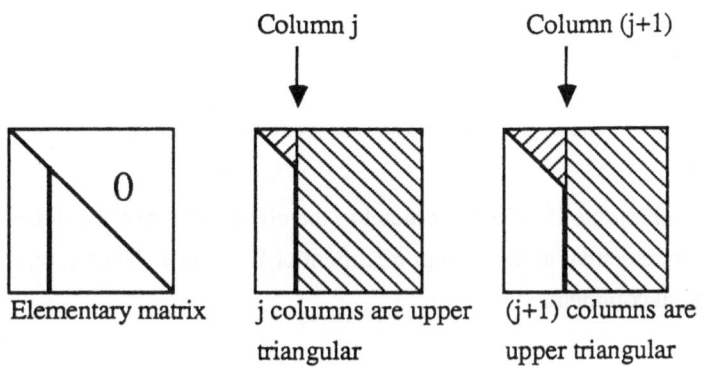

Figure 3.15 A step of Gaussian elimination.

Step 4 Form Etas

From (3.31) and Figure 3.13 the etas of the basis are given by

$$\mathbf{L1 \; L2 \; U2 \; U3} \qquad\qquad (3.32)$$

where

$\mathbf{L1} = \mathbf{L1}_1 \; ... \; \mathbf{L1}_{m1};$

$\mathbf{L2} = \mathbf{L2}_1 \; ... \; \mathbf{L2}_{m2};$

$\mathbf{U2} = \mathbf{U2}_{m2} \; ... \; \mathbf{U2}_1;$

$\mathbf{U3} = \mathbf{U3}_{m3} \; ... \; \mathbf{U3}_1;$

$\mathbf{L1}_j$ is an elementary matrix of size m x m whose j^{th} column is the jth column of $[\mathbf{E1}^t, \mathbf{E4}^t, \mathbf{E2}^t]^t;$

$\mathbf{L2}_j$ is an elementary matrix of size m x m whose $(m1+m3+j)^{th}$ column is the jth column of $[\, \mathbf{0}, \mathbf{0}, \mathbf{L2'}_j^t \,]^t;$

$\mathbf{U2}_j$ is an elementary matrix of size m x m whose $(m1+m3+j)^{th}$ column is the jth column of $[\, \mathbf{0}, \mathbf{G2}, \mathbf{U2'}_j^t \,]^t;$

U3$_j$ is an elementary matrix of size m x m whose (m1+j)th column is the jth column of [**0**, **F2**^{t}, **0**]^t.

Step 5 Update solution :

This is done by copying the original rhs of the given LP and then calling FTRAN to compute the solution **B**$^{-1}$**b**.

3.10.5   Comments

```
C*****************************************************************
C
C    Although Subroutine INVERT is crucial to the effectiveness of RSM, the details of its
C    coding is not essential to the present treatment of DECOMP. Therefore, it is listed below
C    with minimal comments.
C
C*****************************************************************
C                                                          INVERT
         DIMENSION MREG(400), HREG(400), VREG(400)              1
         EQUIVALENCE (MREG(1), YA(1)), (HREG(1), YA(200)), (VREG(1), X(1))  2
C
C        SET PARAMETERS
C
         NELEME = LE(NETA+1) - LE(1)                            3
         WRITE(6, 20) LSUB, NETA, NELEME                       4
   20    FORMAT(/, 3X, 'INVERSION SUBPROBLEM ', I4, 15X, 'OLD   ',   5
        1 I4, ' ETAS , ', I6, ' ELEMENTS')                     6
         CALL TICTAC(JTIM)                                     7
         DFAC = 0.05                                           8
         ZTOLIN = ZTOLZE                                       9
   30    NETA = 0                                             10
         NLETA = 0                                            11
         NGETA = 0                                            12
         NUETA = 0                                            13
         NELEM = LE(1) - 1                                    14
         NLELEM = 0                                           15
         NGELEM = 0                                           16
         NUELEM = 0                                           17
```

```
C                                                    INVERT
         NABOVE = 0                                     18
         LR1 = 1                                        19
         KR1 = 0                                        20
         LR4 = NROW + 1                                 21
         KR4 = NROW                                     22
C
C        PUT SLACKS AND ARTIFICIALS IN PART 4 AND REST IN PART 1
C
         DO 100 I = 1, NROW                             23
         IF (JH(I) .GT. NROW) GO TO 50                  24
         LR4 = LR4 - 1                                  25
         MREG(LR4) = JH(I)                              26
         VREG(LR4) = JH(I)                              27
         GO TO 90                                       28
   50    KR1 = KR1 + 1                                  29
         VREG(KR1) = JH(I)                              30
   90    HREG(I) = -1                                   
                                                        31
         JH(I) = 0                                      32
  100    CONTINUE                                       33
C
         KR3 = LR4 - 1                                  34
         LR3 = LR4                                      35
C
         DO 200 I = LR4, KR4                            36
         IR = MREG(I)                                   37
         HREG(IR) = 0                                   38
         JH(IR) = IR                                    39
         KINBAS(IR) = IR                                40
  200    CONTINUE                                       41
C
C        PULL OUT VECTORS BELOW BUMP AND GET ROW COUNTS
C
         NBNONZ = KR4 - LR4 + 1                         42
         IF (KR1 .EQ. 0) GO TO 1190                     43
         J = LR1                                        44
```

```
C                                                                    INVERT
      210     IV = VREG(J)                                              45
              LL = LA(IV)                                               46
              KK = LA(IV + 1) -1                                        47
              IRCNT = 0                                                 48
              DO 220 I = LL, KK                                         49
              NBNONZ = NBNONZ + 1                                       50
              IR = IA(I)                                                51
              IF (HREG(IR) .GE. 0) GO TO 220                            52
              IRCNT = IRCNT + 1                                         53
              HREG(IR) = HREG(IR) - 1                                   54
              IRP = IR                                                  55
      220     CONTINUE                                                  56
              IF (IRCNT - 1) 230, 250, 300                              57
      230     WRITE(6, 8000) QU, IV, KINBAS(IV)                         58
      8000    FORMAT(' MATRIX SINGULAR  ', A2, 2I6)                     59
              KINBAS(IV) = 0                                            60
              VREG(J) = VREG(KR1)                                       61
              KR1 = KR1 - 1                                             62
              IF (J .GT. KR1) GO TO 310                                 63
              GO TO 210                                                 64
C
      250     VREG(J) = VREG(KR1)                                       65
              KR1 = KR1 - 1                                             66
              LR3 = LR3 - 1                                             67
              VREG(LR3) = IV                                            68
              MREG(LR3) = IRP                                           69
              HREG(IRP) = 0                                             70
              JH(IRP) = IV                                              71
              KINBAS(IV) = IRP                                          72
              IF (J .GT. KR1) GO TO 310                                 73
              GO TO 210                                                 74
      300     IF (J .GE. KR1) GO TO 310                                 75
              J = J + 1                                                 76
              GO TO 210                                                 77
C
C         PULL OUT REMAINING VECTORS ABOVE AND BELOW THE
```

```
C                                                              INVERT
C           BUMP AND ESTABLISH MERIT COUNTS OF COLUMNS
C
   310     NVREM = 0                                              78
           IF(KR1 .EQ. 0) GO TO 1190                              79
           J = LR1                                                80
   320     IV = VREG(J)                                           81
           LL = LA(IV)                                            82
           KK = LA(IV + 1) - 1                                    83
           IRCNT = 0                                              84
           DO 800 I = LL, KK                                      85
           IR = IA(I)                                             86
           IF(HREG(IR) .NE. -2) GO TO 400                         87
C
C          PIVOT ABOVE BUMP (PART OF L)
C
           NABOVE = NABOVE + 1                                    88
           IROWP = IR                                             89
           CALL UNPACK(IV, 2)                                     90
           KVEC = 2                                               91
           CALL WRETA(KVEC)                                       92
           NLETA = NETA                                           93
           JH(IR) = IV                                            94
           KINBAS(IV) = IR                                        95
           VREG(J) = VREG(KR1)                                    96
           KR1 = KR1 - 1                                          97
           NVREM = NVREM + 1                                      98
           HREG(IR) = IV                                          99

           GO TO 940                                             100
C
   400     IF (HREG(IR) .GE. 0) GO TO 800                        101
           IRCNT = IRCNT + 1                                     102
           IRP = IR                                              103
   800     CONTINUE                                              104
C
           IF (IRCNT - 1) 810, 900, 1000                         105
```

```
C                                                              INVERT
  810    WRITE(6, 8000) QR, IV, KINBAS(IV)                       106
         KINBAS(IV) = 0                                          107
         VREG(J) = VREG(KR1)                                     108
         NVREM = NVREM + 1                                       109
         KR1 = KR1 - 1                                           110
         IF (J .GT. KR1) GO TO 1010                              111
         GO TO 320                                               112
C
C        PUT VECTOR BELOW BUMP
C
  900    VREG(J) = VREG(KR1)                                     113
         NVREM = NVREM + 1                                       114
         KR1 = KR1 - 1                                           115
         LR3 = LR3 - 1                                           116
         VREG(LR3) = IV                                          117
         MREG(LR3) = IRP                                         118
         HREG(IRP) = 0                                           119
         JH(IRP) = IV                                            120
         KINBAS(IV) = IRP                                        121
C
C        CHANGE ROW COUNTS
C
  940    DO 950 II = LL, KK                                      122
         IIR = IA(II)                                            123
         IF (HREG(IIR) .GE. 0) GO TO 950                         124
         HREG(IIR) = HREG(IIR) + 1                               125
  950    CONTINUE                                                126
         IF (J .GT. KR1) GO TO 1010                              127
         GO TO 320                                               128
 1000    IF (J .GE. KR1) GO TO 1010                              129
         J = J + 1                                               130
         GO TO 320                                               131
 1010    IF(NVREM .GT. 0) GO TO 310                              132
C
C        GET MERIT COUNTS
C
```

103

```
C                                                        INVERT
 1020    IF (KR1 .EQ. 0) GO TO 1190                         133
         DO 1100 J = LR1, KR1                               134
         IV = VREG(J)                                       135
         LL = LA(IV)                                        136
         KK = LA(IV + 1) - 1                                137
         IMCNT = 0                                          138
         DO 1050 I = LL, KK                                 139
         IR = IA(I)                                         140
         IF (HREG(IR) .GE. 0) GO TO 1050                    141
         IMCNT = IMCNT - (HREG(IR) + 1)                     142
 1050    CONTINUE                                           143
         MREG(J) = IMCNT                                    144
 1100    CONTINUE                                           145
C
C           SORT COLUMNS INTO MERIT ORDER
C            USING SHELL SORT
C
         ISD = 1                                            146
 1106    IF (KR1 .LT. 2*ISD) GO TO 1108                     147
         ISD = 2*ISD                                        148
         GO TO 1106                                         149
 1108    ISD = ISD - 1                                      150
C
C           END OF INITIALIZATION
C
 1101    IF (ISD .LE. 0) GO TO 1107                         151
         ISK = 1                                            152
 1102    ISJ = ISK                                          153
         ISL = ISK + ISD                                    154
         ISY = MREG(ISL)                                    155
         ISZ = VREG(ISL)                                    156
 1103    IF (ISY .LT. MREG(ISJ)) GO TO 1104                 157
 1105    ISL = ISJ + ISD                                    158
         MREG(ISL) = ISY                                    159
         VREG(ISL) = ISZ                                    160
         ISK = ISK + 1                                      161
```

```
C                                                               INVERT
        IF ((ISK + ISD) .LE. KR1) GO TO 1102                    162
        ISD = (ISD - 1) / 2                                     163
        GO TO 1101                                              164
 1104   ISL = ISJ + ISD                                         165
        MREG(ISL) = MREG(ISJ)                                   166
        VREG(ISL) = VREG(ISJ)                                   167
        ISJ = ISJ - ISD                                         168
        IF (ISJ .GT. 0) GO TO 1103                              169
        GO TO 1105                                              170
 1107   CONTINUE                                                171
C
C
C           END OF SORT ROUTINE
C
C       PUT OUT BELOW BUMP ETAS (PART OF U)
C
 1190   NSLCK = 0                                               172
        NBELOW = 0                                              173
        NELAST = NEMAX                                          174
        NTLAST = NTMAX                                          175
        LE(NTLAST + 1) = NELAST + 1                             176
C
        LR = LR3                                                177
        IF (LR3 .GE. LR4) LR = LR4                              178
         IF (LR .GT. KR4) GO TO 2050                            179
        JK = KR4 + 1                                            180
        DO 2000 JJ = LR, KR4                                    181
        JK = JK - 1                                             182
        IV = VREG(JK)                                           183
        I = MREG(JK)                                            184
        NBELOW = NBELOW + 1                                     185
        IF (IV .GT. NROW) GO TO 1200                            186
        NSLCK = NSLCK + 1                                       187
 1200   LL = LA(IV)                                             188
        KK = LA(IV+1) -1                                        189
        IF (KK .GT. LL) GO TO 1300                              190
 1250   IF (DABS(A(LL) - 1.) .LE. ZTOLZE) GO TO 2000            191
```

```
C                                                              INVERT
  1300    NUETA = NUETA + 1                                      192
          DO 1400 J = LL, KK                                     193
          IR = IA(J)                                             194
          IF (IR .EQ. I) GO TO 1390                              195
          IA(NELAST) = IR                                        196
          A(NELAST) = A(J)                                       197
          NELAST = NELAST - 1                                    198
          NUELEM = NUELEM + 1                                    199
          GO TO 1400                                             200
  1390    EP = A(J)                                              201
  1400    CONTINUE                                               202
          IA(NELAST) = I                                         203
          A(NELAST) = 1./EP                                      204
          LE(NTLAST) = NELAST                                    205
          NELAST = NELAST - 1                                    206
          NTLAST = NTLAST - 1                                    207
          NUELEM = NUELEM + 1                                    208
  2000    CONTINUE                                               209
  2050    IF(KR1 .EQ. 0) GO TO 3500                              210
C
C         DO L-U DECOMPOSITION OF BUMP
C
          DO 3000 J = LR1, KR1                                   211
          IV = VREG(J)                                           212
          CALL UNPACK(IV, 2)                                     213
          CALL FTRAN(2, 2)                                       214
          ZTOLR = ZTOLIN                                         215
  2080    DCMAX = 0.0                                            216
          IROWP = 0                                              217
          IRCMIN = -999999                                       218
          DO 2100 I = 1, NROW                                    219
          IF (DABS(YA(I, 2)) .LT. ZTOLR) GO TO 2100              220
          IF (HREG(I) .GE.0) GO TO 2100                          221
           IF(DABS(YA(I, 2)) .GT. DCMAX) DCMAX = DABS(YA(I, 2))  222
          IF (HREG(I) .LE. IRCMIN) GO TO 2100                    223
          IRCMIN = HREG(I)                                       224
```

```
C                                                                    INVERT
      IROWP = I                                                      225
2100  CONTINUE                                                       226
      IF (IROWP .GT. 0) GO TO 2150                                   227
      WRITE(6, 8000) QB, IV, KINBAS(IV)                              228
      KINBAS(IV) = 0                                                 229
      GO TO 3000                                                     230
C
2150  IF(DABS(YA(IROWP, 2)) .GE. DCMAX*DFAC) GO TO 2155             231
      ZTOLR = DCMAX*DFAC                                             232
      GO TO 2080                                                     233
2155  INCR = HREG(IROWP) + 3                                         234
C
C     WRITE L AND U ETAS
C
      IF (J .EQ. KR1) GO TO 2160                                     235
      NELEM = NELEM + 1                                              236
      IA(NELEM) = IROWP                                              237
      A(NELEM) = 1./YA(IROWP, 2)                                     238
      ZTOLET = 0.01*DABS(YA(IROWP, 2))                               239
      IF(ZTOLET .GT. ZTOLZE) ZTOLET = ZTOLZE                         240
2160  DO 2300 I = 1, NROW                                            241
      IF (I .EQ. IROWP) GO TO 2300                                   242
      IF(DABS(YA(I, 2)) .LE. 1.0E-15) GO TO 2300                     243
      IF (HREG(I) .GE. 0) GO TO 2200                                 244
C
C     L ETA ELEMENTS
C
      NELEM = NELEM + 1                                              245
      IA(NELEM) = I                                                  246
      A(NELEM) = YA(I, 2)                                            247
      GO TO 2300                                                     248
C
C     U ETA ELEMENTS
C
2200  IA(NELAST) = I                                                 249
      A(NELAST) = YA(I, 2)                                           250
```

```
C                                                               INVERT
        NELAST = NELAST - 1                                      251
        NUELEM = NUELEM + 1                                      252
 2300   CONTINUE                                                 253
C
        JH(IROWP) = IV                                           254
        KINBAS(IV) = IROWP                                       255
        NUETA = NUETA + 1                                        256
        IA(NELAST) = IROWP                                       257
        IF (J .NE. KR1) GO TO 2330                               258
        A(NELAST) = 1./YA(IROWP, 2)                              259
        GO TO 2340                                               260
 2330   A(NELAST) = 1.                                           261
        NETA = NETA + 1                                          262
        LE(NETA+1) = NELEM + 1                                   263
 2340   NUELEM = NUELEM + 1                                      264
        LE(NTLAST) = NELAST                                      265
        NELAST = NELAST - 1                                      266
        NTLAST = NTLAST - 1                                      267
C
C       UPDATE ROW COUNTS
C
        DO 2350 I = 1, NROW                                      268
        IF (DABS(YA(I, 2)) .LE. 1.0E-15) GO TO 2350             269
        IF (HREG(I) .GE. 0) GO TO 2350                           270
        HREG(I) = HREG(I) - INCR                                 271
        IF (HREG(I) .GE. 0) HREG(I) = -1                         272
 2350   CONTINUE                                                 273
        HREG(IROWP) = 0                                          274
 3000   CONTINUE                                                 275
C
C       SHIFT IE AND E OF U ELEMENTS
C
 3500   NLETA = NETA                                             276
        NETA = NLETA + NUETA                                     277
        NLELEM = NELEM                                           278
        NELEM = NLELEM + NUELEM                                  279
```

```
C                                                                    INVERT
        IF (NUELEM .EQ. 0) GO TO 3550                                   280
        NF = NEMAX - NUELEM + 1                                         281
        INCR = 0                                                        282
        DO 3510 I = NF, NEMAX                                           283
        INCR = INCR + 1                                                 284
        IA(NLELEM + INCR) = IA(I)                                       285
        A(NLELEM + INCR) = A(I)                                         286
3510    CONTINUE                                                        287
        IDIF = NEMAX - NLELEM - NUELEM                                  288
        NF = NTMAX - NUETA + 1                                          289
        INCR = 0                                                        290
        DO 3520 I = NF, NTMAX                                           291
        INCR = INCR + 1                                                 292
        LE(NLETA + INCR) = LE(I) - IDIF                                 293
3520    CONTINUE                                                        294
        LE(NETA+1) = NELEM + 1                                          295
C
C       INSERT SLACKS FOR DELETED COLUMNS
C
3550    DO 3600 I = 1, NROW                                             296
        IF (JH(I) .NE. 0) GO TO 3600                                    297
        JH(I) = I                                                       298
        IROWP = I                                                       299
        CALL UNPACK(I, 2)                                               300
        CALL FTRAN(1, 2)                                                301
        KVEC = 2                                                        302
        CALL WRETA(KVEC)                                                303
3600    CONTINUE                                                        304
C
C       UPDATE X
C
        CALL UNPACK(RHSCOL, 3)                                          305
        IF(LMAX .EQ. 0) GO TO 4100                                      306
        IF(LSUB .GT. 0 ) GO TO 4000                                     307
        DO 3800 I=1, LMAX                                               308
3800    YA((NROWO + I), 3) = 1.                                         309
```

109

```
C                                                              INVERT
       GO TO 4100                                              310
4000   IF(MAST .NE. 2) GO TO 4100                              311
       DO 4050 I = 1, NROWO                                    312
4050   YA(I, 3) = Z(I, 1)                                      313
4100   DO 4200 I = 1, NROW                                     314
4200   YA(I, 2) = YA(I, 3)                                     315
       CALL FTRAN(1, 2)                                        316
       DO 4400 I=1, NROW                                       317
4400   X(I) = YA(I, 2)                                         318
       CALL CHSOL                                              319
       NELEME = LE(NETA+1) - LE(1)                             320
       NSTR = NROW - NSLCK                                     321
       WRITE(6, 5000) ERMAX, NETA, NELEME, NSTR                322
5000   FORMAT(3X, 'MAXIMUM ROW ERROR = ', E14.5, 6X, 'NEW   ', I4, 323
      1 ' ETAS , ', I6, ' ELEMENTS', 4X, I4, ' STRUCTURAL COLUMNS', /)
       IF(ERMAX .LE. 1.E-4) GO TO 6000                         324
       IF(DFAC .GE. .998) GO TO 6000                           325
       DFAC = DFAC*5                                           326
       IF(DFAC .GE. 1) DFAC = .999                             327
       GO TO 30                                                328
6000   CALL TICTAC(KTIM)                                       329
       JTINV=KTIM - JTIM                                       330
C
C      PRINT STATISTICS
C
       NOFD = NELEM - NETA                                     331
       WRITE(6, 8500) NBNONZ, NSTR, NABOVE, NBELOW,            332
      1 NLELEM, NLETA, NUELEM, NUETA, NOFD, NETA
8500   FORMAT(18H0INVERT STATISTICS / 1H , I4, 14H NONZ IN BASIS / 333
      1 1H , I4, 28H STRUCTURAL COLUMNS IN BASIS /
      2 1H , I4, 19H VECTORS ABOVE BUMP /
      3 1H , I4, 19H VECTORS BELOW BUMP /
      4 3H L , I5, 5H NONZ, I5, 5H ETAS/3H U , I5, 5H NONZ, I5, 5H ETAS /
      5 8H TOTALS , I5, 14H OFF DIAG NONZ, I5, 5H ETAS )
99099  RETURN                                                  334
       END                                                     335
```

3.11   Subroutine ITEROP

3.11.1   Major Task

Print simplex iteration log.

3.11.2   Called By

NORMAL

3.11.3   Comments

```
C*********************************************************************
C       IPAR:   Control parameter. '0' directs the printing of headings in output.
C                       '1' directs printing for a major simplex iteration; '3' for a minor iteration.
C       IYY:    Indicates active index in certain arrays.
C
C       Case 1: Print headings.
C               Case 1.1:       Master problem. Print also summary of proposals if any.
C               [lines 27 - 39]
C               Case 1.2:       Subproblem. [lines 22 - 26]
C       Case 2: Print Simplex interation log.
C               Prepare required information. [lines 2 - 10]
C               Case 2.1:       Master problem. [lines 17 - 20]
C               Case 2.2:       Subproblem. [lines 12 - 16]
C
C*********************************************************************
C                                                               ITEROP
        IF (IPAR .EQ. 0) GO TO 1000                             1
        OBJ = - X(IOBJ)                                         2
        IF (IFFEZ .EQ. 0) OBJ = SUMINF                          3
        CALL TICTAC(JTIM)                                       4
        TIMER = (JTIM-ITIM) / 1000.                             5
        CMIN = 0                                                6
        Q = QBL                                                 7
        IF(IPAR .EQ. 1) Q = QM                                  8
        IF(IYY .GE. 1) CMIN = DCMIN(IYY)                        9
        NELEME = LE(NETA+1) - LE(1)                             10
        IF(LSUB .EQ. 0 .AND. MAST .NE. 0) GO TO 500             11
```

111

```
C                                                                              ITEROP
          IF(MAST .EQ. 1 .AND. KFASE .NE. 1) OBJ = DOB + YTEMP(LSUB)      12
C
          WRITE(6, 80) ITCNT, Q, MSTAT, OBJ, IVIN, IVOUT, CMIN, NETA,     13
        1 NELEME, TIMER, NCASUB, QPROP
 80       FORMAT(1H , I5, 1X, A2, 2X, A2, 4X, G16.8, 2I9, 4X, G17.8,      14
        1 2I9, F8.2, 4X, I6, 7X, A2)
          QPROP = QBL                                                     15
          GO TO 9000                                                      16
 500      IF(IVIN .EQ. IOBJ) GO TO 9000                                   17
          WRITE(6, 90) ITCNT, Q, MSTAT, OBJ, IVIN, ICOLS(IVIN), IVOUT,    18
        1 ICOLS(IVOUT), CMIN, NETA, NELEME, TIMER
 90       FORMAT(1H , I5, 1X, A2, 2X, A2, 2X, G16.8, 2(I8, I6), 5X,       19
        1 G16.8, I9, I8, F8.2)
          GO TO 9000                                                      20
1000      IF(MAST .GE. 1 .AND. LSUB .EQ. 0) GO TO 1500                    21
          WRITE(6, 100) LSUB                                              22
 100      FORMAT('0SUBPROBLEM', I5)                                       23
          WRITE(6, 110)                                                   24
 110      FORMAT('0ITCOUNT STATUS   OBJ.VALUE', 8X, 'VECIN', 5X,          25
        1 'VECOUT', 11X, 'DJ', 12X, 'NETA   NELEM    TIME',
        2 5X, 'NO.CAND  CRIT', /)
          GO TO 9000                                                      26
1500      WRITE(6, 140)                                                   27
 140      FORMAT('0RESTRICTED MASTER  PROBLEM', /)                        28
          NUMCAN = NCOL - NCOLO                                           29
          IF(NUMCAN .EQ. 0) GO TO 2000                                    30
          NUMEL = LA(NCOL+1) - LA(NCOLO+1)                                31
          RUMEL = NUMEL                                                   32
          RDENS = RUMEL/(NUMCAN*NROW)*100.                                33
          WRITE(6, 142) NUMCAN, NUMEL                                     34
 142      FORMAT(' TOTAL NUMBER OF CANDIDATES',                           35
        1I5,' WITH', I5, ' ELEMENTS')
          WRITE(6, 145) RDENS                                             36
 145      FORMAT(' MEAN DENSITY OF CANDIDATES ', F6.2, ' PERCENT')        37
2000      WRITE(6, 150)                                                   38
 150      FORMAT('0ITCOUNT STATUS   OBJ.VALUE', 8X, 'VECIN FROM ',        39
```

```
C                                                                    ITEROP
       1   ' VECOUT FROM', 11X, 'DJ', 12X, 'NETA   NELEM   TIME', / )
9000       RETURN                                                      40
           END                                                         41
```

3.12. Subroutine MASTER

3.12.1. Major Task
Performs step 2 of D-W Phase 1 and Phase 2.

3.12.2. Description
MASTER is the driver routine for step 2 of D-W Phase 1 and D-W Phase 2. Occasionally, the decomposition algorithm has to be restarted from a cycle saved in a precious run. In this case, all the data relevant to the problem is read from the unit 9 direct access device (DAD). MASTER then executes a series of cycles till the stopping criterion is met or the maximum limit on the number of cycles has been exceeded. Every cycle begins with solution of the master problem. POLICY is then called provided the previous cycle was not an interrupted cycle (cf. § 2.6.2). Whenever POLICY is called it sets the policy for the current cycle as follows.

1. Sets KK to the number of problems to be solved.
2. Sets the order in which the problems are to be solved.
3. Sets the number of proposals to be generated by each subproblem in the current cycle.

Subsequently, the (KK-1) subproblems identified by POLICY are solved except when a cycle gets interrupted. In this case, assume that the cycle was interrupted while solving the ith subproblem ($1 \leq i \leq$ KK-1). A new cycle is then started by solving the master problem. The first subproblem to be solved in this cycle is either the ith subproblem or the $(i+1)^{st}$ subproblem depending on the strategy chosen by the user. Note that, this cycle may itself be interrupted again in which case the same procedure is repeated. Otherwise, a new cycle would be started by solving the master problem whenever all the remainder of the (KK-1) subproblems have been solved. POLICY is then called and the new cycle is continued by solving subproblems. Master and subproblems are solved as follows.

Case 1 : (Master problem)

First the lower bound of the objective value is computed and then NORMAL is called to solve the master problem. If the maximum cycle limit has been exceeded or if the problem is unbounded then the present run is terminated. Otherwise, the relative difference (RELDIFF) between the primal objective value and the lower bound is computed. If RELDIFF is greater than the tolerance of the stopping criterion (RSTOP) then the solution procedure is continued. The dual variables of the master problem are stored in YA(*, 6) and the appropriate subproblems are solved to generate proposals.

Case 2 : (Subproblem)

The dual variables of the convexity rows of the master problem are stored for performing the candidacy test later on. NORMAL is called to solve the subproblem. The number of proposals generated by the subproblem during the current cycle is recorded. The proposals, if any, are written from the proposal buffer $Z(*, *)$ to the unit 8 work file, provided the user chooses a strategy of solving problems to optimality and sending only q_{max} proposals (cf. § 2.6.2). A check is performed to determine if the cycle should be interrupted and based on this, the next subproblem is solved or a new cycle is started.

3.12.3. Steps

Refer to the psuedo code for MASTER in § 2.7.

3.12.4. Detailed Comments

```
C***********************************************************************
C    Step 1 :                                                   MASTER
C
C    KD        : Points to DAD for accessing the problem's data
C    IRST      : Flag. When set to 1 then the solution procedure is continued from a cycle
C                  saved in a previous run.
C
C    - Initialize variables.  [lines 1 - 9]
C    - If IRST is 1 then set KD so that it points to the DAD for restart information.
C      Read the problem's relevant data.  [lines 11 - 12]
C
C***********************************************************************
```

$$ICPI1 = ICPI + 1 \qquad\qquad 1$$
$$INCO = 0 \qquad\qquad 2$$
$$CALL\ TICTAC(INCIAL) \qquad\qquad 3$$

```
C                                                                    MASTER
            TIMER2 = 0.0                                                4
            NCAND = 0                                                   5
            ICAND = 0                                                   6
            LCOUNT = -1                                                 7
            KK = 1                                                      8
            LS(1) = 0                                                   9
            IF (IRST .EQ. 0) GO TO 200                                 10
            KD = NOLA(1, LMAXP1)                                       11
            READ(9'KD) DOLD, DEOB, TIMER2, NOLA, KCYC,                 12
          1 NCOLO, NROWO, ICOLS, MSTATU
            LCORE = -1                                                 13
            ITEMP = ICPI                                               14
            ICPI = KCYC                                                15
            GO TO 310                                                  16
200         MAJOR = 0                                                  17
            DOLD = -1.0D30                                             18
            GO TO 310                                                  19
300         CALL POLICY(KK, LCOUNT)                                    20
310         IF(KK . LE. 0) GO TO 10000                                 21
            DO 1000 KKK = 1, KK                                        22
315         LSUB = LS(KKK)                                             23
320         IF(LSUB .GE. 1) GO TO 500                                  24
C***************************************************************************
C     SOLVMSTR :
C
C     LS(i)     : Gives the order in which problems are to be solved. Set by POLICY.  E.g. if
C                 LS(2) = 5 then the 2nd problem to be solved in the cycle is problem with
C                 index 5.
C     KK        : Number of problems to be solved before POLICY can be called again.
C     INDEX     : Flag normally set to 1. '0' indicates that the previous cycle was interrupted
C                 while solving the ith subproblem and the first subproblem of the new cycle
C                 is the ith subproblem.
C     NCAND     : Number of proposals made in the current cycle.
C     MSTAT     : Solution status of the problem.
C     NOLA      : Pointer to DAD for accessing a problem's data.
C     DOLD      : Lower bound on the optimal solution.
```

```
C    KCYC    : Cycle number
C    MAJOR   : Cycle number.
C    ICPI    : Cycle number at which special prices are to be read.
C    ICLAST  : Maximum limit on the number of cycles.
C    LMAX    : Number of subproblems.
C    LMAXP1  : Number of subproblems plus 1.
C    ISAVE   : Flag. '1' indicates that the problem is to be saved on DAD for future restarts.
C    YA(*, *) : Contains the master dual variables.
C    KSTR    : Strategy parameter.
C    LCOUNT: Flag used when KSTR is 3.  In this case LCOUNT ≥ -1 indicates that in
C                every cycle after the first, subproblems are to be solved in a cyclic fashion.
C
C         - Print massages.  [lines 34 - 37]
C         - Update the lower bound on the optimal solution as given by eqn (2.75) of
C           § 2.5.3.  [lines 39 - 44]
C         - Call NORMAL to solve the master problem.  [lines 45 - 46]
C         - If the maximum limit on the number of cycles has been exceeded then  save the
C           problem's data on DAD for future restarts if ISAVE is 1.
C           Else proceed.  [lines 47 - 58]
C         - If MSTAT is 'QU' (problem unbounded) then RETURN since the whole problem
C           is unbounded.  [line 60]
C           Else proceed.
C         - Compute the relative difference (RELDIFF) between the objective value and its
C           lower bound and if it is less than RSTOP then STOP.  Else proceed.
C           [lines 64 - 67]
C         - Read the dual variables if the cycle number (MAJOR) is ICPI.  [lines 68 - 71]
C         - Set each subproblem's status feasible and start solving the appropriate
C           subproblem. [lines 78 - 81]
C
C**********************************************************************
C                                                              MASTER
          INDEX = 1                                               25
          IF(NCAND .GE. 1 .OR. LCOUNT .EQ. -1) GO TO 350          26
          GO TO 1000                                              27
350       IVIN = IOBJ                                             28
          MAJOR = MAJOR + 1                                       29
          CALL TICTAC(IFINAL)                                     30
```

```
C                                                                    MASTER
         TIMER1 = (IFINAL - INCIAL) / 1000.                          31
         TIMER2 = TIMER2 + TIMER1                                     32
         INCIAL = IFINAL                                              33
         WRITE(6, 355) MAJOR                                          34
355      FORMAT(////, 40X, 'BEGINNING MAJOR CYCLE ', I4)              35
         WRITE(6, 358) TIMER1, TIMER2                                 36
358      FORMAT(///, 20X, 'TIME FOR PREVIOUS CYCLE = ',               37
        1 F6.2, '  SEC.', 10X, 'TIMIN MASTER  =', F8.2, '  SEC')
         IF (MAJOR .EQ. ICPI1) GO TO 380                              38
         DNEW = 0.                                                    39
         DO 360 I = 1, LULO                                           40
         IF(MSTATU(I) .NE. QBL) GO TO 380                             41
360      DNEW = DNEW + DEOB(I)                                        42
         IF(DNEW .LE. DOLD) GO TO 380                                 43
         DOLD = DNEW                                                  44
380      CALL NORMAL                                                  45
         ICPI = ITEMP                                                 46
         IF (KCYC .NE. ICLAST) GO TO 384                              47
         KCYC = KCYC -1                                               48
         IF (ISAVE .EQ. 0) GO TO 10000                                49
         KD = NOLA(1, LMAXP1)                                         50
         NOLA(2, LMAXP1) = LE(NETA + 1) - 1                           51
         WRITE(9'KD) DOLD, DEOB, TIMER2, NOLA, KCYC,                  52
        1 NCOLO, NROWO, ICOLS, MSTATU
         KD = NOLA(3, LMAXP1)                                         53
         IAUX = NOLA(2, LMAXP1)                                       54
         KIKA = 1                                                     55
         CALL VECTOR(KIKA, A, IA, NB, IAUX)                           56
         WRITE (6, 6020)                                              57
6020     FORMAT( // '0PROBLEM SAVED FOR RESTART')                     58
         GO TO 10000                                                  59
384      IF(MSTAT .EQ. QU) GO TO 10000                                60
         DEOB(LULO) = -X(IOBJ)                                        61
         MSTATU(LULO) = MSTAT                                         62
         IF(MSTAT .NE. QBL .OR. DOLD .LE. -1.E10) GO TO 388           63
         DNEW = (DEOB(LULO) - DOLD) / DABS(X(IOBJ))                   64
```

```
C                                                                 MASTER
         WRITE(6, 385) DOLD, DNEW                                    65
385      FORMAT(//, 3X, 'LOWER BOUND  = ', F16.8, // 1H ,            66
        1 'RELATIVE DIFFERENCE = ', F16.8)
         IF(DNEW .LT. RSTOP) GO TO 10000                            67
388      IF (MAJOR .EQ. ICPI) READ(5, 5000) (YA(I, 6), I = 1, NROW) 68
5000     FORMAT(6G12.6)                                             69
         WRITE (6, 6000)                                            70
6000     FORMAT(/ '0MASTER DUAL VARIABLES' /)                       71
         WRITE (6, 6010) ((I, YA(I, 6)), I = 1, NROW)               72
6010     FORMAT(6(1X, I3, 1X, G12.6))                               73
         IF(LMAX .EQ. 0) GO TO 1000                                 74
         IF(INCO .NE. -1) GO TO 390                                 75
         INCO = 0                                                   76
390      IF(ITCNT .EQ. ITMA .AND. ITCNT .NE. 0) GO TO 1000          77
         DO 400 I = 1, LMAX                                         78
400      MSTATU(I) = QF                                             79
         IF(INDEX .EQ. 0) GO TO 315                                 80
         GO TO 1000                                                 81
```

```
C************************************************************************
C     SOLVSUB :
C
C     YTEMP(i): Contains dual variable of the ith convexity constraint.
C     LSUB    : Problem index of the subproblem being solved.
C     NCASUB: Number of proposals made by the subproblem being solved.
C     ICASUB(i): Number of proposals made by the ith subproblem in the current cycle.
C     MAXCA   : Maximum number of proposals that can be made by a subproblem in a cycle.
C     NCAND   : Number of proposals generated in the current cycle.
C     MAXNCA: Once NCAND ≥ MAXNCA the current cycle is interrupted.
C     ICAND   : Pointer to NCAND for the index of the last proposal generated by the
C                 previous subproblem.
C     KSTR    : Strategy parameter. '5' indicates that the subproblem is to be solved to
C                 optimality and only the last MAXCA proposals are sent.
C                 '4' indicates that a subproblem is allowed to generate up to MAXCA
C                 proposals. If the subproblem is still not optimized a new cycle is begun with
C                 the same subproblem as the first subproblem.
```

```
C    INDEX   : Flag used when KSTR is 4.  In this case, a cycle is interrupted when INDEX
C                 is set to 0.
C    MSTAT   : Status of the problem.
C    DEOB(i) : Objective value for subproblem i minus the dual variable of the ith convexity
C                 constraint.
C
C            - Variables are initialized.  [lines 84 - 88]
C            - The dual variable of the convexity constraint is saved in YTEMP(*).
C              [lines 82 - 83]
C            - NORMAL is called to solve the subproblem and generate proposals. [line 89]
C            - The number of proposals generated by the subproblem is recorded. [line 90]
C            - If there are any unbounded proposals then to discover other unbounded columns,
C              NORMAL would have set MSTAT to 'QF' (feasible).  Reset MSTAT to 'QU'
C              (unbounded).
C            - If KSTR is '5' then the proposals, if any, are written on an unit 8 work  file.
C              [lines 97 - 100]
C            - If the subproblem is infeasible then STOP since the whole problem is infeasible.
C              [line 101]
C              Otherwise proceed.
C            - The current cycle would be interrupted if KSTR is '4' and the subproblem is not
C              optimized. Then INDEX is set to zero.  [lines 106, 108]
C              A new cycle is started by solving the master problem.  [lines 109 - 110]
C              Since INDEX is zero, on completion of the master problem, the subproblem last
C              solved will be the first subproblem of the new cycle.  [line 80]
C            - If the number of proposals generated in the current cycle is ≥ MAXNCA the cycle
C               is interrupted except if the current subproblem is the last subproblem.  [line 107]
C
C
C******************************************************************************
C                                                                     MASTER
 500      DO 520 I = 1, LMAX                                             82
 520      YTEMP(I) = YA(NROWO + I, 6)                                    83
          NCASUB = 0                                                     84
          DOBMA = 0.0                                                    85
          ISUNB = 0                                                      86
          NCUNB = 0                                                      87
          ITSLCA = -5                                                    88
```

```
C                                                               MASTER
          CALL NORMAL                                             89
          ICASUB(LSUB) = NCASUB                                   90
          IF (ISUNB .LE. 0) GO TO 530                             91
C
          DO 525 I = 1, ISUNB                                     92
          KINBAS(ICUNB(I)) = 0                                    93
  525     ICUNB(I) = 0                                            94
C
          MSTAT = QU                                              95
  530     CONTINUE                                                96
          IF (KSTR .NE. 5 .OR. NCASUB .EQ. 0) GO TO 550           97
          IPROS = 0                                               98
          WRITE (8) ((Z(I, J), I=1, NZMAX), J = 1, NCASUB)        99
          ICAND = ICAND + NCASUB                                 100
  550     IF (MSTAT .EQ. QN) GO TO 10000                         101
          MSTATU(LSUB) = MSTAT                                   102
          DEOB(LSUB) = DOB + YTEMP(LSUB)                         103
          DO 620 I = 1, LMAX                                     104
  620     YA((NROWO + I), 6) = YTEMP(I)                          105
          IF(KSTR .EQ. 4 .AND. MSTAT .NE. QBL) GO TO 700         106
          IF(NCAND.LT.MAXNCA .OR. KKK. EQ. (KK - 1)) GO TO 1000  107
  700     IF(KSTR .EQ. 4 .AND. MSTAT .NE. QBL) INDEX = 0         108
          LSUB = 0                                               109
          GO TO 350                                              110
 1000     CONTINUE                                               111
          GO TO 300                                              112
10000     IF (LMAX .EQ. 0) GO TO 9999                            113
          WRITE(6, 6990) IPROP                                   114
 6990     FORMAT('1PROPOSAL GENERATION STATISTICS'/              115
          '0CRITERION', 10X, 'R    F    A    B    O    U'/        116
          ' NO PROPOSALS   ', 6I5)                               117
 9999     IF (ISOL .NE. 1) STOP                                  118
          RETURN                                                 119
          END                                                    120
```

3.13. Subroutine NORMAL

3.13.1. Major Task

Solves an LP using RSM (cf. § 2.3).

3.13.2. Inputs

The index number of the problem (master or subproblem) to be solved.

3.13.3. Description

NORMAL is called by INDATA or MASTER whenever a problem has to be solved. Since it is not possible to store all the problems in core at the same time, all the data relevant to the problem are stored on a unit 9 direct access device (DAD). Whenever a problem has to be solved NORMAL calls CHANGE. If the problem is not in core then CHANGE first writes the problem currently in core to DAD and reads the problem that needs to be solved from DAD. Also, if there is no storage available in the eta file, CHANGE sets the number of etas (NETA) to zero. Once the problem is in core NORMAL calls INVERT if

(a) the number of etas NETA is zero, or

(b) the infinity norm of the residual rhs exceeds 10^{-4} (cf. § 3.4). The infinity norm is computed by calling CHSOL.

If the problem being solved is the master problem, then PACK is called to incorporate the proposals into the data matrix. Also, since solving the master problem implies the beginning of a new cycle, one should check to ensure that the maximum cycle has not been exceeded before proceeding further. A set of candidate columns can then be selected for multiple pricing as follows:

Case 1 : (Solving master problem) The cost vector is formed in the buffer YA(*, 6) by calling FORMC. Since proposals by definition have favorable reduced costs, a set of columns is selected in PACK as the proposals get incorporated. Thereby, a full scale pricing operation is avoided. However, the user can specify a cycle number at which a complete pricing is desired to start solving the master problem. In this case PRICE is called following BTRAN to select columns for multiple pricing.

Case 2 : (Solving subproblem in D-W Phase 1 or D-W Phase 3) The cost vector is formed in YA(*, 6) by calling FORMC. Then BTRAN and PRICE are called to select columns for multiple pricing.

Case 3 : (Solving subproblems in D-W Phase 2) In this case the subproblem's feasible cost vector (cf. § 2.6.1) is :

$$[1, -\pi_0{}^k, 0] \tag{3.33}$$

and the corresponding dual vector is

$$[1, -\pi_0{}^k, -\pi_r{}^k] \tag{3.34}$$

Reference to subroutine MASTER (cf §3.12) will show that at the beginning of a major cycle the master sets up the cost vector given by (3.33) in a buffer named YA(*, 6). If during the course of a major cycle, a subproblem's solution becomes infeasible, then to set up the cost vector for infeasibilities FORMC would set YA(*, 6) to the zero vector. Hence, the cost vector would not be available for solving subsequent subproblems. To avoid this difficulty, a parameter KFASE is maintained in D-W Phase 2 to indicate the status of the subproblem's solution and the cost vector is set up accordingly. When a subproblem's solution is infeasible in D-W Phase 2 then KFASE is set to 1 and the cost vector is set up in YA(*, 5). Otherwise KFASE is set to 2 and the cost vector is formed in YA(*, 6). The sequence of calls to select columns for multiple pricing in case 3 is as follows:

(a) Initialize KFASE to 1 and call FORMC.
(b) If the solution is infeasible then the cost vector would have been formed in YA(*, 5) since KFASE is 1. Go to (d).
(c) If the solution is feasible then KFASE is set to 2 and FORMC is called again to set the third component $-\pi_r{}^k$ to the zero vector.
(d) Call PRICE following BTRAN to select columns for multiple pricing.

If no columns can be selected then the problem is infeasible if the current solution is infeasible and optimal otherwise. In the latter case CHECK is called to see if a proposal can be made. On the other hand, if columns have been selected then multiple pricing can be started.

Multiple pricing is begun by unpacking the selected columns in a buffer YA and the following is done to each one of those columns. FTRAN is first called to update each candidate column. CHUZR then determines

(a) the allowable increase if the variable corresponding to the candidate column enters;
(b) the resulting change in the objective value; and
(c) whether the column is unbounded.

Case 1 : (CHUZR identifies the column as unbounded) If the current problem is the master problem then the whole problem is unbounded. If the problem is in step 1 of D-W Phase 1 then the proposal will be written to the proposal buffer Z(*, *) in the subroutine INDATA. If the problem is in D-W Phase 3 then a proposal need not be formed. Otherwise CHECK is called to ascertain the possibility of making a proposal. CHECK may or may not be able to make a proposal without exceeding the maximum limit on the number of proposals that can be generated by a subproblem. In the latter case the solution of the subproblem is interrupted and control is returned to NORMAL. In the former case, the column is fixed at zero and the optimization of the restricted subproblem is continued. This is done (cf. § 2.6.2) to enable other extreme rays to be discovered.

Case 2 : (Not an unbounded column) The best change in the objective value is updated. This procedure is repeated for the each column in the candidate pool stored in the buffer YA(*, *).

Therefore, at the end of one pass through all candidate columns in YA(*, *), the one that provided the best change DY in the objective value is selected to be the entering column provided none of the columns were unbounded. To find the pivot row, CHUZR is called again to pick the maximum pivot among those columns which provide the same change DY in the objective value. The solution is updated and a basis change is effected by marking the leaving column as non-basic and the entering column as basic. Also, the column that just entered the basis is marked to prevent it from being considered for entering the basis during the next multiple pricing pass. CHECK is called again and two cases may result.

Case 1 : (Do not continue with multiple pricing) A proposal can be generated provided some tests (cf. § 2.6.2) are passed by a subproblem in D-W Phase 2 and a strategy of generating intermediate proposals (cf. § 2.6.2 and § 5.2) is chosen by the user. If the maximum limit on the number of proposals has been exceeded then the solution of the problem has to be terminated. CHECK sets a flag KOUT to signal this condition. Often, an INVERT may be necessary to improve on the accuracy of a proposal. In this case it is better to interrupt the multiple pricing cycle and start from the beginning of the simplex cycle. CHECK sets the flag IRTN to denote this condition. If both KOUT and IRTN have not been set then multiple pricing is continued as in Case 2 after a proposal is made.

Case 2 : (Continue with multiple pricing) Conceivably the dual prices after the basis change performed before the call to CHECK would not be very different from that of the previous basis. Therefore, all columns in YA(*, *) except the ones which have already entered the basis may still price out favorably. To verify this, one has to update the column with the lastest eta and compute the new reduced costs. This is done efficiently by calling FTRAN with the parameter IPAR set to '3' and the next multiple pricing pass can be started. If all the columns selected in the buffer YA(*, *) have either unfavorable reduced costs or have already been pivoted in, then a fresh simplex iteration is started.

3.13.4. Steps

1. Read the problem from DAD.

2. Factorize the basis if necessary.

3. Pack proposals if necessary and select a pool of columns for multiple pricing. If no columns can be selected, check for optimality or unboundedness and go to step 6.

4. Perform multiple pricing.

5. Go to step 2 to begin the next simplex iteration.

6. If the problem is optimal then call CHECK. RETURN.

3.13.5. Deatailed Comments

```
C*******************************************************************
C     Step 1 :                                              NORMAL
C
C     LSUP   : Index of the problem to be solved.
C
C     (i)    Set LSUP to index of problem to be solved. [line 6]
C            Initialize other variables. [lines 1 - 5, 8 - 11]
C     (ii)   Call CHANGE to write to DAD the problem currently in-core and
C            read the LSUP^th problem into core from DAD. [line 7]
C
C*******************************************************************
```

IPAR = 1	1
ITEMP = 0	2
INVF = INVFRQ	3
IF (LSUB .EQ. 0) INVF = INVFMA	4
CALL TICTAC(ITIM)	5
LSUP = LSUB	6
CALL CHANGE(LSUP)	7
KFASE = 2	8
NCASUB = 0	9
JTINV = 0	10
ITMA = ITCNT	11

```
C*******************************************************************
C     Step 2 :
C
C     NETA      : Number of etas.
C     ERMAX     : Residual rhs ∂b.
C     YA(*, 3)  : Contains b to start with and contains ∂b after CHSOL.
```

```
C    NROWO       : Number of coupling rows.
C
C    (i)    If NETA is 0 then call INVERT.  [lines 12, 30].
C           Otherwise, if residual rhs ≥ 10⁻⁴ then call INVERT. The residual rhs ∂b is
C           computed by CHSOL which needs as inputs the rhs and the computed solution
C           x+∂x (cf. (3.10) in  § 3.4).  These inputs are provided as follows.
C           -  The rhs supplied by the user is stored in YA(*, 3) by calling UNPACK.
C              [line 13]
C              If it is the master problem then 1's are inserted for the convexity rows.
C              [lines 16 - 18]
C              If the program is executing D-W Phase 3 then the allocation for the subproblem,
C              computed by (2.74), is stored in the buffer Z(*, 1).  This is copied to the
C              appropriate locations in YA. [lines 20 - 22]
C            -  Call FTRAN to compute x+∂x and copy it to the array X.  [lines 22 - 26]
C    (ii)   Call CHSOL and then INVERT if necessary.  [lines 27 - 28]
C
C***********************************************************************
C                                                                   NORMAL
              IF(NETA .EQ. 0) GO TO 1000                               12
              CALL UNPACK(RHSCOL, 3)                                   13
              IF(LMAX .EQ. 0) GO TO 4100                               14
              IF(LSUB .GT. 0 ) GO TO 4000                              15
              DO 3800 I = 1, LMAX                                      16
     3800     YA((NROWO + I), 3) = 1.                                  17
              GO TO 4100                                               18
     4000     IF(MAST .NE. 2) GO TO 4100                               19
              DO 4050 I = 1, NROWO                                     20
     4050     YA(I, 3) = Z(I, 1)                                       21
     4100     DO 4200 I = 1, NROW                                      22
     4200     YA(I, 2) = YA(I, 3)                                      23
              CALL FTRAN(1, 2)                                         24
              DO 4400 I = 1, NROW                                      25
     4400     X(I) = YA(I, 2)                                          26
              CALL CHSOL                                               27
              IF(ERMAX .LE. 1.E-4) GO TO 1060                          28
     1000     NT = NT + JINTV                                          29
              CALL INVERT                                              30
```

C   NORMAL
        ITSINV = 0   31

C**
C   Step 3 :
C
C   MAST   : Indicates the stage of the problem. '0' indicates step 1 of D-W phase 1.
C   '1' indicates step 2 of D-W Phase 1 or D-W Phase 2. '2' indicates D-W
C   Phase 3.
C   NCAND : Number of proposals generated in the current cycle.
C   KCYC : Cycle number.
C   ICLAST : Maximum cycle limit.
C   KFASE : Normally set to '2' for the cost vector to be formed in YA(*, 6).
C   When a subproblem's solution becomes infeasible in D-W Phase 2
C   then KFASE is set to '1' for the cost vector to be formed in YA(*, 5).
C   MSTAT : Solution status of the problem.
C
C   (i)   If there are proposals and the master problem is being solved then call PACK to
C   incorporate the proposals into the data array A. [lines 33, 36 - 37]
C   (ii)   If the cycle number has exceeded the limit, then RETURN. [lines 38, 42]
C   (iii) Candidate columns for multiple pricing are selected as follows:
C   *Case 1* : (Master problem)
C   - Call FORMC to form the cost vector in YA(*, 6)
C   - Candidate columns for multiple pricing are selected by PACK except if special
C   prices are to be computed. In this case BTRAN and PRICE are called.
C   [lines 46 - 47, 58 - 59]
C   *Case 2* : (Subproblem in D-W Phase 1 or Phase 3)
C   - Call FORMC, BTRAN, PRICE. [line 55 - 56, 58 - 59]
C   *Case 3* : (Subproblem in D-W Phase 2)
C   - Refer to § 3.13.3.
C   (a) [lines 48 - 49]
C   (b) [line 50]
C   (c) [lines 51 - 55]
C   (d) [lines 58 - 59]
C   (iv) If no candidate columns are selected by multiple pricing and the current solution is
C   infeasible then set MSTAT to 'QN' and RETURN.   [lines 60 - 61]
C   If no candidate columns are selected by multiple pricing and the current solution is

126

```
C        feasible then the problem is optimal.  Hence, set MSTAT to 'QBL' and call
C        CHECK to see if a proposal can be made.  [lines 60 - 63, 138]
C        Else go to Step 4.
C
C*************************************************************************
C                                                                 NORMAL
         IF(ITCNT .NE. ITMA) GO TO 1070                                32
1060     IF(LSUB .EQ. 0 .AND. NCAND .GE. 1) GO TO 1070                 33
         CALL ITEROP(0, IYIV)                                          34
1070     IF(MAST .NE. 1) GO TO 1500                                    35
         IF(LSUB .GE. 1) GO TO 1100                                    36
         IF(NCAND .NE. 0) GO TO 1075                                   37
         IF (KCYC .LT. ICLAST) GO TO 1500                              38
         RETURN                                                        39
1075     CALL PACK                                                     40
         NCAND = 0                                                     41
         IF (KCYC .LT. ICLAST) GO TO 1080                              42
         RETURN                                                        43
1080     CALL ITEROP(0, IYIV)                                          44
         CALL FORMC                                                    45
         IF (KCYC .EQ. ICPI) GO TO 1800                                46
         GO TO 1900                                                    47
1100     KFASE = 1                                                     48
         CALL FORMC                                                    49
         IF(MSTAT .EQ. QI) GO TO 1800                                  50
1200     KFASE = 2                                                     51
         DOB = 0.                                                      52
         DO 1400 I = 1, NROWO                                          53
1400     DOB = DOB - X(I)*YA(I, 6)                                     54
1500     CALL FORMC                                                    55
         IF(KFASE .NE. 1) GO TO 1800                                   56
         IF(MSTAT .EQ. QF) GO TO 1200                                  57
1800     CALL BTRAN                                                    58
         CALL PRICE                                                    59
1900     IF(MULT .GT. 0) GO TO 3000                                    60
         IF (MSTAT . EQ. QI ) GO TO 2000                               61
         MSTAT = QBL                                                   62
```

127

```
C                                                                  NORMAL
          GO TO 5810                                                   63
2000      MSTAT = QN                                                   64
          GO TO 6000                                                   65


C****************************************************************
C     Step 4 :
C
C     YA(*, *)      : Buffer containing candidate columns in multiple pricing.
C     JCOLP(i)      : Column index in the data matrix A of the column YA(*, i).
C     LSUB          : Index number of the problem being solved.
C     MAST          : Indicates the stage of the problem.
C     MAXCA         : Maximum limit on the number of proposals that can be made by a
C                     subproblem in a cycle.
C     KSTR          : Strategy parameter. '5' indicates that only the last maxca proposals are to
C                     be sent to the master problem.
C     NCASUB        : Number of proposals generated by the LSUB^th subproblem in the
C                     current cycle.
C     KVEC          : Column of YA that provides the best change among all the candidates.
C     MSTAT         : Status of the problem.
C     KTYPE         : Ratio type.  (cf. subroutine CHUZR)
C     NTYPE         : Ratio type.  (cf. subroutine CHUZR)
C     DE            : Allowable increase in the variable corresponding to the candidate
C                     column.
C     DY            : Allowable change in the objective value if the variable corresponding to
C                     the candidate column enters.
C     IROWP         : Pivot row for the candidate column that provides the best change in the
C                     objective value.
C     IPAR          : Parameter indicating type of FTRAN
C     JENTRY        : Parameter indicating type of CHUZR.
C     KOUT          : Flag set by CHECK to indicate if the solution of the problem is to be
C                     terminated for the current cycle.
C     IRTN          : Flag set by CHECK to indicate if the multiple pricing cycle is to be
C                     interrupted.
C
C     (i)    Unpack all candidate columns selected for multiple pricing in the buffer YA.
C            [lines 66 - 68]
```

C   (ii)   If the candidate column has already entered the basis (KINBAS(j) ≠ 0) then it

C   cannot enter the basis again. Hence, take up the next column. Otherwise call

C   FTRAN to update the candidate column being considered. [lines 74 - 76]

C   (iii) If the reduced cost of the column is not favorable then it cannot enter. Hence, take

C   up the next column. Otherwise call CHUZR to determine the allowable increase if

C   the candidate column enters the basis. CHUZR also computes the resulting change

C   in the objective value and find out if the column is unbounded. In addition CHUZR

C   finds the pivot row if the column were to enter. [lines 77 - 78]

C

C   *Case 1* : (Unbounded Column identified)

C   - Call CHECK if a subproblem is being solved in D-W Phase 2. [lines 79 - 82]

C   - If a proposal is made without exceeding the limit on the number of proposals

C   generated by a subproblem then reset MSTAT to QF and fix the column at zero.

C   Interrupt the multiple pricing cycle and start a new simplex iteration. This is done

C   to discover other extreme ray proposals if any. [lines 83 - 92]

C   Otherwise RETURN.

C   *Case 2* : (Not an unbounded column)

C   - Update the best change in the objective value. [lines 95 - 96]

C   - Update the column index of the candidate column that yielded this change in the

C   objective value. [line 99]

C   - Update the allowable change if the variable corresponding to the candidate column

C   enters. [lines 100]

C   (iv) Take up the next column till all columns in YA have been considered.

C   (v)   Set KVEC to the column of YA providing the best change in the objective value.

C   [line 104].

C   If the ratio type (KTYPE) is not '2' then call CHUZR with JENTRY set to '2' to

C   find the maximum pivot row. [lines 105, 108]

C   Update the solution by calling UPBETA. [line 111]

C   (vi) Update the basis. [lines 112 - 117]

C   (vii) Print iteration log. [line 119]

C   (viii) Call CHECK to see if a proposal can be made. [line 118]

C   *Case 1* : (Do not continue with multiple pricing)

C   *Case 1.1* : (terminate the solution of the problem)

C   - If KOUT is 1 then the solution of the problem is to be terminated for the

C   current cycle. [lines 120, 129]

C   - If there is space in the eta file then write the etas. [lines 129 - 130]

C   *Case 1.2* : (Terminate multiple pricing)

129

```
C                      - If IRTN is 1 then an INVERT was done in CHECK.
C                        Hence, start a new simplex cycle.  [lines 120 - 121]
C                Case 2 :  (Continue with multiple pricing)
C                - Call WRETA to write the etas generated after the last basis update.  [line 123]
C                - Set IPAR to 3 so that the reduced cost will also be computed.  [line 125]
C      (ix)  Go to (ii) to take up the remaining candidate columns of YA.
C      (x)   If all columns in YA do not have favorable reduced cost or have already
C            entered the basis then go to Step 2.  [lines 134 - 136]
C
C****************************************************************************
C                                                                    NORMAL
3000      DO 3200 I = 1, MULT                                        66
          J = JCOLP(I)                                              67
 3200     CALL UNPACK(J, I)                                         68
          IPAR = 1                                                  69
          DO 5000 K = 1, MULT                                       70
          DELTA = 1.5                                               71
          DO 4500 I = 1, MULT                                       72
          ITEMP = I                                                 73
          IF(JCOLP(I) .EQ. 0) GO TO 4500                            74
          IF(KINBAS(JCOLP(I)) .NE. 0) GO TO 4500                    75
          CALL FTRAN(IPAR, I)                                       76
          IF(DCMIN(I) .GE. -ZTCOST) GO TO 4500                      77
          CALL CHUZR(I, NTYPE, 1)                                   78
          IF (MSTAT .NE. QU) GO TO 4460                             79
          IF (LSUB .EQ. 0) GO TO 5800                               80
          IF (MAST .NE. 1) GO TO 5800                               81
          CALL CHECK(KOUT, ITEMP, IRTN)                             82
          IF(KSTR .EQ. 5 .AND. NCUNB .EQ. MAXCA) GO TO 4450         83
          IF(KSTR .NE. 5 .AND. NCASUB .GE. NC(KKK)) GO TO 4450      84
          IF(ISUNB .GE. 7) GO TO 4450                               85
          ISUNB = ISUNB + 1                                         86
          ICUNB(ISUNB) = JCOLP(I)                                   87
          KINBAS(JCOLP(I)) = -1                                     88
          JCOLP(I) = 0                                              89
          MSTAT = QF                                                90
          IF (IRTN .EQ. 1) GO TO 1800                               91
```

```
C                                                           NORMAL
          GO TO 1500                                           92
4450      IYIV = ITEMP                                         93
          GO TO 6000                                           94
4460      IF(DELTA .LE. DY) GO TO 4500                         95
          DELTA = DY                                           96
          KTYPE = NTYPE                                        97
          MROWP = IROWP                                        98
          IYIV = I                                             99
          DTEMP = DE                                          100
4500      CONTINUE                                            101
          IF(DELTA .GT. 0.1) GO TO 5200                       102
          DE = DTEMP                                          103
          KVEC = IYIV                                         104
          IF(KTYPE .NE. 2) GO TO 4600                         105
          IROWP = MROWP                                       106
          GO TO 4700                                          107
4600      CALL CHUZR(KVEC, KTYPE, 2)                          108
4700      IVIN = JCOLP(IYIV)                                  109
          IVOUT = JH(IROWP)                                   110
          CALL UPBETA(IYIV)                                   111
          KINBAS(IVIN) = IROWP                                112
          KINBAS(IVOUT) = 0                                   113
          JH(IROWP) = IVIN                                    114
          ITCNT = ITCNT + 1                                   115
          ITSINV = ITSINV + 1                                 116
          ITEMP = IYIV                                        117
          CALL CHECK(KOUT, ITEMP, IRTN)                       118
          CALL ITEROP(IPAR, IYIV)                             119
          IF(KOUT .EQ. 1) GO TO 4800                          120
          IF (IRTN .EQ. 1) GO TO 1800                         121
          IF(NELEM .GT. (NEMAX - NROW)) GO TO 1000            122
          CALL WRETA(IYIV)                                    123
          JCOLP(IYIV) = 0                                     124
          IF(IPAR .EQ. 1) IPAR = 3                            125
          GO TO 5000                                          126
4800      IF (NELEM .GT. NEMAX - NROW) GO TO 4850             127
```

C		NORMAL
	IF (IRTN .EQ.1) GO TO 6010	128
	CALL WRETA(IYIV)	129
	GO TO 6010	130
4850	NETA = 0	131
	GO TO 6010	132
5000	CONTINUE	133
5200	IF(NELEM .GT. (NEMAX - NROW)) GO TO 1000	134
	TIMER = JTINV	135
	IF(ITSINV - INVF) 1500, 1000, 1000	136
5800	IYIV = ITEMP	137
5810	CALL CHECK(KOUT, ITEMP, IRTN)	138
6000	CALL ITEROP(IPAR, IYIV)	139
6010	NT = NT + JTINV	140
	RETURN	141
	END	142

3.14.   Subroutine PACK

3.14.1. Major Task

Packs a proposal into the master problem data array and finds the reduced costs of various proposal columns.

3.14.2. Inputs

1. Proposal buffer Z(*, *).
2. The unit 8 work file storing proposals that cannot be accommodated in Z(*, *).

3.14.3. Called By

INDATA and NORMAL.

3.14.4. Description

In order to solve the master problem, all proposal columns generated in the current cycle must be incorporated into the data array A(*) of the master problem. Proposals are copied one at a time from a proposal buffer known as the Z array. Let NPROS be the maximum number of columns that can be accommodated by Z. This number depends on the strategy chosen. When there are more than NPROS proposals, Z contains only those generated last. The others have been written on the unit 8 work file. To process the proposals, the current contents of Z are first written to the end of the unit 8 work file. Then the first NPROS proposals are read into Z from the begining of the work file. After these proposals are incorporated, the next NPROS proposals are read from the work file and so on until all proposals have been processed. The actual incorporation of a proposal is explained below.

Before a proposal is incorporated, one should check for availability of space in the array A. If there is not enough room, then some of the "inactive" columns in A have to be purged. A column is labelled as "active" if it is basic or if it is non-basic with reduced cost "better" that a threshold value or if it has been generated in the current cycle. Purging involves scanning through all columns in A to determine if a certain column is "active" or not. An inactive column in A is purged by overwriting it with the next active column as illustrated in Figure 3.16. If the space created by the current pass of purging is not enough, then the threshold reduced cost used to qualify active columns is tightened further and the process is repeated.

When there is space to pack the proposal, it is scaled if it is an extreme point proposal and then incorporated into the data array $A(*)$ of the master problem.

If the maximum number of columns that can be chosen for multiple pricing (cf. § 2.2.4) has not been exceeded, then the current column is selected for multiple pricing provided its reduced cost is "better" than zero. Otherwise it is chosen if its reduced cost is "better" than the "worst" reduced cost for columns already chosen for multiple pricing.

3.14.5. Steps

    1. Set up the Z-buffer.

    2. For each proposal do

        (a) Check if purging is required and if so go to step 2b. Else go to step 2c.

        (b) Purge if necessary.

        (c) If the contents of the current Z-buffer have been processed then read next batch of columns from the work file.

        (d) Scale the column if necessary and incorporate it into the data array.

        (e) Select columns for multiple pricing.

    End For.

Column (j-1)   Column j

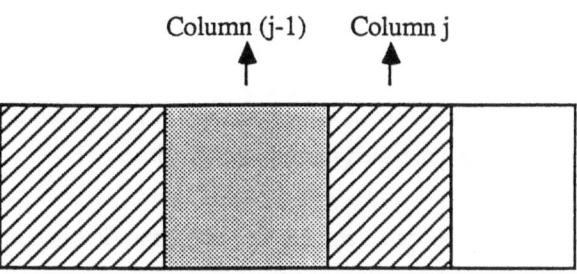

i. Column (j-1) is to be purged.

ii.Column j is written over Column (j-1).

Column j

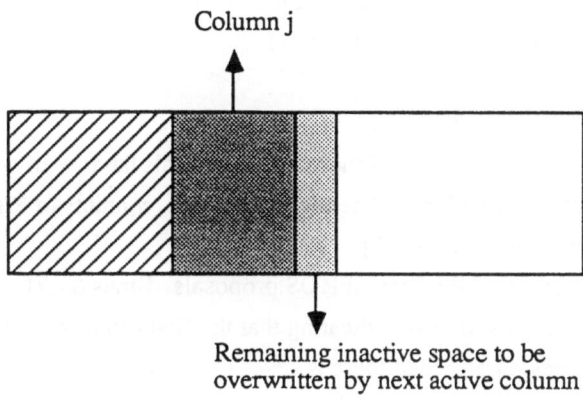

Remaining inactive space to be
overwritten by next active column

Figure 3.16 Data array for master problem: before and after purging a proposal column

3.14.6. Detailed Comments

```
C*********************************************************************
C     Step 1 :                                                  PACK
C
C     KSTR :   '5' indicates that a partial cycle strategy is adopted. In this case a major cycle
C              is interrupted once the total number of proposals exceeds the supplied limit.
C              The next major cycle is started by solving the next  subproblem. For
C              example,  if the limit is exceeded while solving the ith subproblem, then the
C              next cycle starts with subproblem i+1. Therefore subproblems 1 through i
C              would not be generating any proposal in this new cycle.
C
C     MAST:    '0' implies that the program is executing step 1 of  D-W Phase 1 and during
C              this phase each subproblem generates only one proposal.
```

134

```
C
C       Case 1 : (KSTR = 5 and MAST ≠ 0)
C                   In this case a cycle interruption could have occured and some problems
C                   would have generated more than one proposal.
C
C               - Scan through all problems to see which one first generated proposals after a
C                   cycle interruption. Let it be the L^th problem. [lines 11 - 13]
C               - Set JEND to the number of proposals generated by thr L^th problem. [line 14]
C               - Read JEND proposals into Z. [line 15]
C               - Set ICASUB(L) to zero to ensure that the next time the proposal from the
C                   same problem is not read again. [line 16]
C               - Set IPROS to zero indicating that the first proposal in Z can now be processed
C                   if there is space in A. [line 17]
C
C       Case 2 : ( KSTR ≠ 5 or MAST = 0 )
C               - If NCAND > NPROS then more proposals have been generated than can be
C                   accommodated in Z. Therefore, write the contents of Z to the end of the unit 8
C                   work file. [lines 4 - 5]
C               - Read into Z the first NPROS proposals. [lines 6 - 7]
C               - Set IPROS to zero, indicating that the first proposal in Z can now be
C                   processed provided there is space in A. [line 17]
C
C**************************************************************************
C                                                                    PACK
        MULT = 0                                                        1
        DCOST = - ZTCOST                                                2
        IF (KSTR .EQ. 5 .AND. MAST .NE. 0) GO TO 90                     3
        IF (NCAND .LE. NPROS) GO TO 100                                 4
        WRITE (8) Z                                                     5
        REWIND 8                                                        6
        READ (8) Z                                                      7
        GO TO 100                                                       8
90      L = 1                                                           9
        REWIND 8                                                       10
92      IF (ICASUB(L) .NE. 0) GO TO 94                                 11
        L = L+ 1                                                       12
        GO TO 92                                                       13
```

```
C                                                                    PACK
 94       JEND = ICASUB(L)                                            14
          READ (8) ((Z(I, J), I = 1, NZMAX), J = 1, JEND)            15
          ICASUB(L) = 0                                               16
100       IPROS = 0                                                   17
          DO 2000 K = 1, NCAND                                       `18


C***********************************************************************
C     Step 2a :
C
C     - If there is no space in the data array A(*)  or if the inclusion of the proposal means
C       exceeding the maximum number of columns then go to step 2b to purge inactive
C       columns.  Else proceed.  [line 19]
C     - Increment IPROS since one more column can be processed.  [line 20]
C
C***********************************************************************
          IF ((LA(NCOL+1) .GT. LE(1) - NROW) .OR.                    19
         1(NCOL+1 .GT. NLAMAX)) GO TO 1000
200       IPROS = IPROS + 1                                          20


C***********************************************************************
C     Step 2c :
C
C     Case 1 :  (KSTR = 5 and MAST ≠ 0)
C        - If all proposals in Z have not been processed (IPROS ≤ JEND) then go to step 2d
C          to process the proposal currently being pointed at in Z.
C        - If IPROS > JEND then all proposals in Z have been processed.  Read into Z the
C          proposals from the problem that next generated proposals.  [lines 22 - 28]
C
C     Case 2 :  ( KSTR ≠ 5 or MAST = 0 )
C        - If IPROS  < NPROS then all proposals in Z have been incorporated.  Read into Z
C          the next NPROS proposals.  [lines 29 - 30]
C
C***********************************************************************
          IF (KSTR .NE. 5 .OR. MAST .NE. 1) GO TO 208                21
          IF (IPROS .LE. JEND) GO TO 210                             22
```

```
C                                                                    PACK
202     L = L + 1                                                     23
        IF (ICASUB(L) .EQ. 0) GO TO 202                               24
        JEND = ICASUB(L)                                              25
        ICASUB(L) = 0                                                 26
        READ(8) ((Z(I, J), I = 1, NZMAX), J = 1, JEND)                27
        GO TO 209                                                     28
208     IF (IPROS .LE. NPROS) GO TO 210                               29
        READ (8) Z                                                    30
209     IPROS = 1                                                     31


C********************************************************************
C     Step 2d :
C
C     Case 1 :  (Extreme point proposal, indicated by NAME(*) > 0)
C          (a)  The 2-norm of the proposal vector is found.  [lines 34 - 36]
C               Using this 2-norm the scaling factor DSY is computed.  [lines 38 - 39]
C          (b)  Each element in the proposal is first scaled and tested to see if it is not  zero.
C               If non-zero, NELA, the number of non-zero elements in A is incremented and
C               the element is incorporated into the data array A(*).  [lines 48 - 50]
C          (c)  The origin of the proposal is recorded (i.e. the subproblem that generated it) so
C               that one can keep track of the number of proposals generated by individual
C               subproblems.  [line 55]
C          (d)  Since it is an extreme point proposal there should be a '1' in the convexity row.
C               Incorporate the scaled value of 1 into A.   [lines 56 - 59]
C          (e)  In CHECK, Z(NROW+1, *) was set to the reduced cost of the column.  Set
C               DSUM to the scaled value of the reduced cost so that a decision can be made in
C               step 2e in testing the column for multiple pricing.  [line 61]
C
C     Case 2 :  (extreme ray proposal, indicated by NAME(*) ≤ 0)
C          (a)  In this case scaling is not done and the value of the scaling factor DSY is 1.
C          (b)  Each non-zero entry in the proposal is incorporated as explained in case 1.
C               [lines 43 - 51]
C          (c)  Record the origin of the proposal.  [line 53]
C          (d)  Same as (e) of case1.
C
C********************************************************************
```

```
C                                                                    PACK
210       DSUM = 1.                                                    32
          IF(NAME(K) .LE. 0) GO TO 312                                 33
          DO 300 I = 1, NROWO                                          34
300       DSUM = DSUM + Z(I, IPROS)*Z(I, IPROS)                        35
          DSUM = DSQRT(DSUM)                                           36
          IF (NAME(K) .LE. 0 ) GO TO 312                               37
          DSY = 10. / DSUM                                             38
          IF (DSUM .GT. 1000.) DSY = 0.01                             39
          GO TO 315                                                    40
312       DSY = 1.                                                     41
          IF (DSUM .LT. 1.) DSY = 10./DSUM                             42
315       NELA = LA(NCOL + 1) - 1                                      43
          DO 500 I = 1, NROWO                                          44
          DSA = Z(I, IPROS)*DSY                                        45
          IF(DABS(DSA) .LE. 1.E-15) GO TO 500                          46
          NELA = NELA + 1                                              47
          IA(NELA) = I                                                 48
          A(NELA) = DSA                                                49
500       CONTINUE                                                     50
          NCOL = NCOL + 1                                              51
          IF(NAME(K) .GE. 0) GO TO 600                                 52
          ICOLS(NCOL) = -NAME(K)                                       53
          GO TO 800                                                    54
600       ICOLS(NCOL) = NAME(K)                                        55
          NELA = NELA + 1                                              56
          IA(NELA) = NROWO + NAME(K)                                   57
          A(NELA) = DSY                                                58
800       LA(NCOL + 1) = NELA + 1                                      59
          KINBAS(NCOL) = 0                                             60
          DSUM = Z(NROWO + 1, IPROS)*DSY                               61


C*****************************************************************
C     Step 2e :
C
C     DCOST    : Threshold reduced cost. Initialized to zero tolerance.
C     DSUM     : The reduced cost of the column just processed.
```

```
C     MULT     : The number of columns selected for multiple pricing.
C     KMULT    : Set by the user and indicates the maximum number of columns that can be
C                  selected for multiple pricing.
C     DCMIN    : Array containing the reduced cost of the columns selected for multiple
C                  pricing.
C     JMIN     : Index of the column with the "worst" reduced cost among those selected
C                  for multiple pricing.
C     JCOLP(i) : Column index of the i-th column selected for multiple pricing.
C
C     Case 1  : (More columns may be selected)
C         - Check to see if the reduced cost DSUM is < 0 and if so proceed.  [line 62]
C           This is done since, as long as less than (KMULT) columns have been selected,
C           any column with reduced cost < 0 is eligible for multiple pricing. Increment MULT
C           and select this column for multiple pricing.  [lines 72 - 75]
C
C     Case 2  : (KMULT columns have been selected)
C         - In this case a threshold reduced cost of 0 is not good enough since there are
C           KMULT columns with reduced costs "better" than 0. Therefore the threshold
C           reduced cost DCOST is updated to the value of the "worst" reduced cost from
C           among the set of columns selected for multiple pricing.  [lines 66 - 71]
C         - A column is selected only if its reduced cost is "better" than the previously updated
C           threshold value DCOST and it replaces the column with the worst reduced cost.
C           [lines 61 - 65]
C
C*********************************************************************
C                                                                   PACK
          IF(DSUM .GE. DCOST) GO TO 2000                             62
          IF(MULT .LT. KMULT) GO TO 850                              63
          DCMIN(JMIN) = DSUM                                         64
          JCOLP(JMIN) = NCOL                                         65
  820     DCOST = DSUM                                               66
          DO 830 J = 1, KMULT                                        67
          IF(DCOST .GE. DCMIN(J)) GO TO 830                          68
          DCOST = DCMIN(J)                                           69
          JMIN = J                                                   70
  830     CONTINUE                                                   71
          GO TO 2000                                                 72
```

```
C                                                              PACK
850      MULT = MULT + 1                                        73
         DCMIN(MULT) = DSUM                                     74
         JCOLP(MULT) = NCOL                                     75
         IF(MULT .LT. KMULT) GO TO 2000                         76
         JMIN = MULT                                            77
         GO TO 820                                              78
```

```
C**********************************************************************
C     Step 2b :
C
C     DE    : Threshold reduced cost. Initialized to 10.
C     KK    : The total number of columns in A before a new round of purging begins.
C             Since there are NCOL columns to start with, KK is initialized to NCOL.
C     NCOL  : The number of "active" columns in A at any point.
C             Since all coupling rows are active, NCOL is initialized to the number of
C             coupling rows NCOLO. [line 82]
C     NELA  : The number of elements in A that are active. NELA is initialized to the
C              number of elements in the coupling columns. [lines 83]
C
C     For each column from NCOLO+1 to KK :
C        Case 1 : (Active column not to be purged)
C        - If the column is basic; [line 88]
C          or if it is generated in the current cycle; [line 89]
C          or if its reduced cost is "better" than the threshold value; [lines 90 - 93]
C          then it is active and should not be purged.
C        - Starting from NELA write the active column over an inactive column. NELA is
C          updated to indicate the number of active elements in A. [lines 94 - 97]
C        - Since one more column has been added, increment the number of columns NCOL.
C          [lines 98]
C        - Since the column index and the number of elements of a column are changed due to
C          purging, update LA, ICOLS, KINBAS and JH. [lines 98 - 102]
C        - If the column is selected for multiple pricing update the column index in JCOLP.
C          [lines 104 - 107]
C
C        Case 2 : (Inactive column to be purged)
C             - A column can be purged if it does not pass the tests described in case 1.
```

```
C               It may be overwritten with an active column.
C      End For.
C               If the current round of purging is not sufficient then tighten the reduced cost
C      threshold and start a new round of purging. [lines 111 - 113]
C
C*********************************************************************
C                                                            PACK
1000   DE = 10.                                               79
       DY = LA(NCOL + 1) - LA(NCOLO + 1)                      80
1050   KK = NCOL                                              81
       NCOL = NCOLO                                           82
       NELA = LA(NCOLO + 1) - 1                               83
       MM = NCOLO + 1                                         84
       DO 1500 I = MM, KK                                     85
       MU = LA(I + 1) - 1                                     86
       NN = LA(I)                                             87
       IF(KINBAS(I) .NE. 0) GO TO 1150                        88
       IF(I .GT. (KK + 1 - K)) GO TO 1150                     89
       DP = 0.                                                90
       DO 1100 J = NN, MU                                     91
1100   DP = DP + A(J)*YA(IA(J), 6)                            92
       IF(DP .GE. DE) GO TO 1500                              93
1150   DO 1200 J = NN, MU                                     94
       NELA = NELA + 1                                        95
       IA(NELA) = IA(J)                                       96
1200   A(NELA) = A(J)                                         97
C
       NCOL = NCOL + 1                                        98
       LA(NCOL + 1) = NELA + 1                                99
       ICOLS(NCOL) = ICOLS(I)                                 100
       KINBAS(NCOL) = KINBAS(I)                               101
       IF(KINBAS(NCOL) .GT. 0) JH(KINBAS(NCOL)) = NCOL        102
       IF(I .LE. (KK - K + 1)) GO TO 1500                     103
       IF(MULT .EQ. 0) GO TO 1500                             104
       DO 1400 JA = 1, MULT                                   105
       IF(JCOLP(JA) .NE. I) GO TO 1400                        106
       JCOLP(JA) = NCOL                                       107
```

```
C                                                                   PACK
        GO TO 1500                                                  108
1400    CONTINUE                                                    109
1500    CONTINUE                                                    110
        DP = (LA(NCOL+1) - LA(NCOLO+1))/DY                          111
        IF(NELA .LT. NEMAX .AND. DP .LT. 0.75) GO TO 200            112
        DE = 0.1*DE - ZTCOST                                        113
        GO TO 1050                                                  114
2000    CONTINUE                                                    115
        IF (NCAND .GT. NPROS .OR. KSTR .EQ. 5) REWIND 8             116
        IPROS = 0                                                   117
        ICAND = 0                                                   118
        RETURN                                                      119
        END                                                         120
```

3.15. Subroutine POLICY

3.15.1. Major Task
Sets proposal generation policy.

3.15.2. Outputs
Sets policy if necessary. This involves :

1. Setting limts on the maximum number of proposals that can be generated by a subproblem in a cycle.

2. Setting the order in which subproblems are to be solved.

3.15.3. Called By
Called by MASTER for each uninterrupted cycle.

3.15.4. Description
POLICY first checks to see if a policy has to be set as follows. If there are no subproblems then policy needs not be set. If all subproblems are optimal then the policy that was set during a previous call is still valid. Otherwise policy is set according to the value of the control parameter KSTR.

3.15.5. Steps

1. Decide if policy has to be set and if so go to step 2. Else RETURN.
2. Set policy.

3.15.6. Detailed Comments

```
C**********************************************************************
C     Step 1 :
C
C            LMAX     : Maximum number of subproblems.
C            MSTAT(*): Status of the *th subproblem.  QBL indicates optimal,  and QN
C                         indicates infeasible.
C            KK       : Maximum number of problems to be solved in a cycle.
C            LS(*)    : Index of the *th subproblem to be solved. Zero indicates master
C                         problem.
C
C     (i)    If LMAX is 0 then there are no subproblems.  Therefore, if the master is optimal
C            or infeasible then we are done.  Otherwise, KK is set to 1 and LS(1) is set to zero
C            indicating that only the master problem is to be solved.  If LMAX > 0 then
C            proceed. [lines 2, 8 - 12]
C     (ii)   If all problems are optimal then policy need not be set.  Otherwise go to
C            step 2.  [lines 3 - 7]
C**********************************************************************
C                                                      .                       POLICY
        KK = -1                                                                   1
        IF(LMAX .EQ. 0) GO TO 500                                                 2
        DO 200 I = 1, LMAX                                                        3
        IF(MSTATU(I) .EQ. QBL) GO TO 200                                          4
        GO TO 990                                                                 5
  200   CONTINUE                                                                  6
        GO TO 10000                                                               7
  500   IF(MSTAT .EQ. QBL .OR. MSTAT .EQ. QN) GO TO 10000                         8
        KK = 1                                                                    9
        LS(1) = 0                                                                10
        KRIT(1) = 1                                                              11
        GO TO 10000                                                             12


C**********************************************************************
```

```
C    Step 2 :                                                    POLICY
C
C    LMAX   : Maximum number of subproblems.
C    LS(*)  : Index of the *th subproblem to be solved. Zero indicates master problem.
C    NC(*)  : Number of proposals to be generated by the *th subproblem.
C    KRIT(*) : Indicates stopping criterion when solving subproblem LS(*).  '1' indicates
C             that one proposal is to be generated at optimality or unboundedness.
C             '2' means that NC(*) proposals are to be generated.
C    KK     : Maximum number of subproblems to be solved in a cycle.
C    MAXCA : Maximum number of proposals that can be generated in a cycle by any
C             subproblem.
C    KSTR   : Control parameter based on which policy is set.
C    LCOUNT: Used when KSTR is '3'.  Its exact significance will be explained later.
C
C         Case 1 :  (KSTR = 1)
C         In this case all subproblems are solved and one proposal is to be generated at
C    optimality or unboundedness.  Therefore, LS(i) is i, NC(i) is '1' and KRIT(i) is '1'.
C    [lines 15 - 17]
C         Since all subproblems are solved, a total of LMAX+1 problems are
C    solved (including master) and the (LMAX+1)st problem is the master
C    problem.  [lines 9 - 10]
C         Case 2 :  (KSTR = 2)
C         In this case every subproblem is solved and a maximum of MAXCA proposals
C    can be generated in a cycle by any subproblem.  Therefore, LS(i) is i, NC(i) is set to
C    MAXCA and KRIT(i) is set to '2'.  [lines 25 - 27]
C         Again a total of LMAX+1 problems are solved (KK is set to LMAX+1) and the
C    last problem is the master problem.  [lines 28 - 30]
C         Case 3 :  (KSTR = 3)
C         In this case, during the first cycle, every subproblem is solved and a maximum
C    of MAXCA proposals can be generated by any subproblem.  Subsequently, only one
C    subproblem per cycle is solved in a cyclic manner. Consider an example with five
C    subproblems. If the fifth subproblem was solved in a certain cycle then in the following
C    cycle only the first subproblem is solved.  The variable LCOUNT is kept to indicate the
C    subproblem(s) to be solved in a cycle.  '-1' indicates the first cycle and all subproblems
C    are solved to generate a maximum of MAXCA proposals.  Hence when LCOUNT is -1
C    it is just like the case when KSTR is 2.  [lines 32, 22]  Otherwise, only the LCOUNTth
C    subproblem is solved during a cycle.  [lines 36 - 40]
```

```
C        To maintain the cyclic order of solving the subproblems, LCOUNT is updated to
C   (LCOUNT mod LMAX) at the end of every cycle. [lines 33 - 34]
C
C        Case 4 : (KSTR = 4 and 5).
C        As far as POLICY is concerned this is the same as KSTR = 2.  [lines 13]
C
C****************************************************************************
C                                                                    POLICY
990     GO TO (1000, 2000, 3000, 2000, 2000), KSTR                      13
1000    DO 1200 I = 1, LMAX                                             14
        LS(I) = I                                                      15
        KRIT(I) = 1                                                    16
1200    NC(I) = 1                                                      17
        LS(LMAX + 1) = 0                                               18
        KRIT(LMAX + 1) = 1                                             19
        KK = LMAX + 1                                                  20
        GO TO 10000                                                    21
1900    LCOUNT = LCOUNT + 1                                            22
2000    DO 2200 I = 1, LMAX                                            23
        ICASUB(I) = 0                                                  24
        LS(I) = I                                                      25
        KRIT(I) = 2                                                    26
2200    NC(I) = MAXCA                                                  27
        LS(LMAX + 1) = 0                                               28
        KRIT(LMAX + 1) = 1                                             29
        KK = LMAX + 1                                                  30
        GO TO 10000                                                    31
3000    IF(LCOUNT .EQ. -1) GO TO 1900                                  32
3100    LCOUNT = LCOUNT + 1                                            33
        IF(LCOUNT .GT. LMAX) LCOUNT = 1                                34
        IF(MSTATU(LCOUNT) .EQ. QBL) GO TO 3100                        35
        LS(1) = LCOUNT                                                 36
        KRIT(1) = 2                                                    37
        NC(1) = MAXCA                                                  38
        LS(2) = 0                                                      39
        KRIT(2) = 1                                                    40
        KK = 2                                                         41
```

C   POLICY
10000 RETURN   42
        END   43

3.16. Subroutine PRICE

3.16.1. Major Task

    Selects columns for multiple pricing.

3.16.2. Input

    1. The array A(*).

    2. The dual prices in the buffer YA(*, KVEC).

    3. KMULT indicating the maximum number of columns that can be selected for multiple pricing. This is a control parameter specified by the user.

3.16.3. Outputs

    1. Sets MULT to the number of columns selected for multiple pricing.

    2. The column index of the columns selected for multiple pricing is stored in JCOLP(*).

    3. The reduced costs of each of these columns is stored in DCMIN(*).

3.16.4. Called By

    NORMAL

3.16.5. Description

    PRICE computes the reduced cost of each nonbasic column and determines whether the column should be selected for multiple pricing. The reduced cost of the column a_j (cf. § 2.2.3) is computed using πa_j where π is the dual vector.

    A column can be chosen for multiple pricing provided its reduced cost is "better" than a threshold value. The threshold value is initially set to zero and is updated according to the following cases.

    Case 1: (Number of columns chosen < KMULT). Since the quota for columns has not yet been filled, any column with negative reduced cost is eligible. Therefore, the threshold reduced cost remains at zero.

    Case 2 : (Number of columns selected = KMULT). In this case the threshold reduced cost is updated to the value of the "worst" reduced cost of columns currently chosen for multiple pricing. A "better" column replaces the column with the "worst" reduced cost and so keeping the number of columns selected for multiple pricing to the user specified limit of KMULT.

As a column is selected for multiple pricing, its column index in the matrix **A** is stored in the array JCOLP(*). Its reduced costs is stored in the array DCMIN(*).

3.16.6. Steps

1. Initialize.
2. For each column i do
      (a) compute the reduced cost;
      (b) decide if the column should be selected for multiple pricing.
   End for.

3.16.7. Detailed Comments

```
C**********************************************************************
C     Step 1 :                                                  PRICE
C
C     MULT   : Number of columns selected for multiple pricing.
C     DCOST  : Threshold value of the reduced cost.
C     ZTCOST : Zero tolerance for reduced cost.
C     KVEC   : Column index of the buffer YA(*, *) that contains the dual vector π.
C              When KFASE is '1' the dual vector is in YA(*, 5). Otherwise it is in
C              YA(*, 6).
C
C**********************************************************************
          IYIV = 0                                                    1
          MULT = 0                                                    2
          DCOST = -ZTCOST                                             3
          KVEC = 6                                                    4
          IF(KFASE .EQ. 1) KVEC = 5                                   5


C**********************************************************************
C     Step 2a :
C
C     KINBAS(*)      : '0' indicates that the *th column is nonbasic.
C     YA(*, KVEC)    : Contains the dual vector.
```

```
C    A(*)             : Contains the columns to be priced.
C    DSUM             : Holds inner product in computing reduced cost.
C    LL               : Position in A(*) of the first non-zero of the column being
C                       considered.
C    KK               : Position in A(*) of the last non-zero of the column being
C                       considered.
C    ISTYPE(*)        : Indicates the type of the *th row.  '-1' indicates an equality row;
C                       '1' indicates an inequality row.
C
C    (i)     Ignore basic columns.  Also ignore equality logicals since they are not allowed
C            to enter the basis.  [lines 7 - 9]
C    (ii)    Compute the inner product πaj which gives the reduced cost.  [lines 11 - 14]
C
C*******************************************************************
C                                                              PRICE
         DO 1000 J = 1, NCOL                                      6
         IF (J .GT. NROW) GO TO 499                               7
         IF (ISTYPE(J) .NE. 1) GO TO 1000                         8
499      IF (KINBAS(J) .NE. 0) GO TO 1000                         9
         DSUM = 0.                                               10
         LL = LA(J)                                              11
         KK = LA(J + 1) - 1                                      12
         DO 500 I = LL, KK                                       13
500      DSUM = DSUM + YA(IA(I), KVEC)*A(I)                      14


C*******************************************************************
C    Step 2b :
C
C        DSUM      : The reduced cost of the column being considered.
C        DCOST     : Threshold value of the reduced cost.  Initialized to zero.
C        JCOLP(*)  : Index of the *th column selected for multiple pricing.
C        DCMIN(*)  : Reduced cost of the *th column selected for multiple pricing.
C        MULT      : Number of columns selected for multiple pricing.
C
C    (i)     If the reduced cost of the column being considered is greater than the threshold
C            value then it cannot be selected for multiple pricing.  Hence go to step 2a.
C            Otherwise there are two cases.
```

148

```
C              Case 1 : (MULT < KMULT)
C                  Increment MULT to indicate one more column selected for multiple
C              pricing.  Store the reduced cost and the column index of the column.
C              [lines 26 - 28]
C                  If the column just selected increases MULT to KMULT then the threshold
C              reduced cost is updated.  [lines 29 - 31, 19 - 24]
C                  Case 2 : (MULT = KMULT)
C                  Update the value of the threshold reduced cost.  [lines 19 - 24]
C
C*********************************************************************
C                                                              PRICE
               IF(DSUM .GE. DCOST) GO TO 1000                     15
               IF(MULT .LT. KMULT) GO TO 700                      16
               DCMIN(JMIN) = DSUM                                 17
               JCOLP(JMIN) = J                                    18
      550      DCOST = DSUM                                       19
               DO 600 I = 1, KMULT                                20
               IF(DCOST .GE. DCMIN(I)) GO TO 600                  21
               DCOST = DCMIN(I)                                   22
               JMIN = I                                           23
      600      CONTINUE                                           24
               GO TO 1000                                         25
      700      MULT = MULT + 1                                    26
               DCMIN(MULT) = DSUM                                 27
               JCOLP(MULT) = J                                    28
               IF(MULT .LT. KMULT) GO TO 1000                     29
               JMIN = MULT                                        30
               GO TO 550                                          31
      1000     CONTINUE                                           32
               RETURN                                             33
               END                                                34
```

3.17. Subroutine RESULT

3.17.1. Major Task
Reconstructs the primal solution vector.

3.17.2. Calling
FORMC, UNRAVL, BTRAN, INVERT, NORMAL, and UNPACK.

3.17.3. Called By
Main Program.

3.17.4. Description
RESULT first checks if the master problem or one of the subproblems is infeasible. In this case Phase 3 is abandoned. Otherwise, the feasible vector to be reconstructed (cf § 2.5.3) is :

$$(x_0{}^*, x_1{}^*, \ldots , x_R{}^*) \qquad (3.35)$$

First, the vector $x_0{}^*$ is obtained from the second component of the solution to the master problem (cf. § 2.5.3). Then variable names are read from DAD (direct access device) and the subroutine UNRAVL outputs the vector $x_0{}^*$.

Next, each $x_r{}^*$ ($r = 1,..,R$) is reconstructed by solving the subproblem (cf § 2.5.3):

$$
\begin{aligned}
\text{Minimize} \quad & c_r x_r \\
\text{subject to} \quad & A1_r x_r = y_r \\
& A2_r x_r = b_r \qquad\qquad (3.36)\\
& x_r \geq 0
\end{aligned}
$$

where $A1_r$ is m_0 by n_r and $A2_r$ is m_r by n_r and all other vectors and matrices are of suitable dimensions. This is done in two steps. The first step is to compute y_r and the second step is to actually solve the subproblem (3.36) defined above.

1. Computing y_r : The vector y_r, known as the allocation vector, contains the allocation of each coupling resource to subproblem r. Consider an example with three coupling rows and two subproblems. Assume that the problem was optimal after two cycles and that each subproblem generated one proposal per cycle. Let $q^k{}_{ij}$ denote the jth element of the kth

proposal generated by subproblem i. Let v_i and u_i indicate the weight on the ith proposal from subproblems 1 and 2 respectively. Then, the constraints of the master problem at the end of two cycles are given in (3.37) below:

$$s_1 \qquad \mathbf{a1}_{01}x_0 + q_{11}{}^1v_1 + q_{11}{}^2v_2 + q_{21}{}^1w_1 + q_{21}{}^2w_2 = d_{01}$$

$$s_2 \qquad \mathbf{a1}_{02}x_0 + q_{12}{}^1v_1 + q_{12}{}^2v_2 + q_{22}{}^1w_1 + q_{22}{}^2w_2 = d_{02}$$

$$s_3 \qquad \mathbf{a1}_{03}x_0 + q_{13}{}^1v_1 + q_{13}{}^2v_2 + q_{23}{}^1w_1 + q_{23}{}^2w_2 = d_{03} \qquad (3.37)$$

$$s_4 \qquad\qquad + v_1 \quad + v_2 \qquad\qquad\qquad = 1$$

$$s_5 \qquad\qquad\qquad\qquad\qquad + w_1 \quad + w_2 \quad = 1$$

where s_i denotes the logical for the ith constraint, $\mathbf{b}_{0i}$ denotes the right-hand side (resource availability) of the ith coupling constraint and $\mathbf{a1}_{0i}$ denotes the ith row of the coupling columns $\mathbf{A1}_0$, (cf § 2.5.3) of the problem. The allocation vector $\mathbf{y}_r$ ($r = 1, 2$) is

$$\mathbf{y}_1 = (q^1{}_{11}v_1, q^1{}_{12}v_1, q^1{}_{13}v_1)^t + (q^2{}_{11}v_2, q^2{}_{12}v_2, q^2{}_{13}v_2)^t \qquad (3.38)$$

for subproblem 1 and

$$\mathbf{y}_2 = (q_{21}{}^1w_1, q_{22}{}^1w_1, q_{23}{}^1w_1)^t + (q_{21}{}^2w_2, q_{22}{}^2w_2, q_{23}{}^2w_2)^t \qquad (3.39)$$

for subproblem 2. For each subproblem r, $\mathbf{y}_r$ is computed and written to the unit 8 work file. To compute $\mathbf{y}_r$, a scan is performed through all rows (including the convexity row) of the master problem. For each row, the column that is basic in that row is determined. If the column was generated by the rth subproblem then the allocation vector $\mathbf{y}_r$ is updated by adding the contribution from the column. Otherwise, the next row is taken up.

After computing $\mathbf{y}_r$, an array named XRHS (i) is initialized to :

$$\mathbf{b}_{0i} - \mathbf{a1}_{0i}\, x_0 \qquad \text{for all } i = 1,..., m_0 \qquad (3.40)$$

where $\mathbf{a1}_{0i}$ is the ith row of $\mathbf{A1}_0$. Therefore, at this stage, XRHS denotes the amount of resource left after subtracting the resource consumed by the master problem.

2. Solving subproblem : At the end of D-W Phase 2, let π_r^* be the dual vector corresponding to the constraints $A2_r x_r = b_r$. Without loss of generality, the reconstruction of x_r^* can be stated as follows :

2.1. The subproblem is brought into memory by calling CHANGE and reading the allocation vector y_r from the unit 8 work file. INVERT is then called to begin the solution of the subproblem.

2.2. The dual vector π_r^*, which is to be part of the output, is not readily available at this stage. Also, the dual vector $\pi'_r{}^*$ for the constraints $A2_r x_r = b_r$ obtained by solving (3.36) would not in general be the same as π_r^*. Therefore, FORMC and BTRAN are called to recompute π_r^* before going to step 2.3.

2.3. The type of the rows corresponding to the constraints $A1_r x_r = y_r$ is changed from "non-binding" to "equality". NORMAL is called to solve the problem (3.36).

2.4. The array XRHS is updated so that after treating subproblem r, it contains :

$$b_{0i} \; - \; \sum_{j=0}^{r} A1_{ji} \, x_j \qquad i = 1,..., m_0 \qquad (3.41)$$

In other words, after solving the rth subproblem the amount of resources consumed by that subproblem is subtracted from what was left after treating subproblems 1 through (r-1).

2.5. UNRAVL is called to output the solution.

3.17.5. Steps

1. Check if Phase 3 is to be abandoned. If so STOP. Else proceed.

2. Compute resource allocations.

3. Solve Phase 3 subproblems.

3.17.6. Detailed Comments

```
C******************************************************************
C     Step 1 :                                            RESULT
C
C         MSTAT    : Solution status of the problem.
C         LCORE    : Flag to indicate if the LCORE^th problem is in core.
```

```
C          LSUB      : Index of the problem being solved. '0' indicates the master
C                       problem.
C          LMAXP1    : Number of subproblems plus one.
C          NOLA(4, LSUB): Pointer to the direct access device (DAD) where the names
C                       (of rows and columns) of the LSUB^th problem are stored.
C
C     - Phase 3 is abandoned if :
C          - one of the subproblems is infeasible [line 1, 6 - 13];  or
C          - the master problem is either unbounded or infeasible.  [lines 2, 4, 14 - 16]
C     Else proceed.  [line 3]
C          - Solve the master problem if the solution is not already in core.   [lines 17 - 19]
C          - Read master problem's names.  [lines 20 - 21]
C          - Call UNRAVL to output x_0*.  [line 22]
C
C*********************************************************************************
C                                                                      RESULT
          IF(MSTAT .EQ. QN .AND. LCORE .GE. 1) GO TO 1000            1
          IF(MSTAT .EQ. QU) GO TO 1500                               2
          IF(MSTAT .EQ. QBL) GO TO 2000                              3
          IF(MSTAT .EQ. QN) GO TO 1000                               4
          STOP 1234                                                  5
1000      WRITE(6, 1200) LSUB                                        6
1200      FORMAT(1H1, 'SUBPROBLEM ', I4, '   HAS NO SOLUTION')       7
          MM = LSUB                                                  8
          IF(LSUB .EQ. 0) MM = LMAXP1                                9
          KD = NOLA(4, MM)                                          10
          READ(9'KD) NAME(1), NAME(2), ICNAM                        11
          CALL UNRAVL                                               12
          GO TO 10000                                               13
1500      WRITE(6, 1600)                                            14
1600      FORMAT(1H1, ' MASTER PROBLEM IS UNBOUNDED')               15
          GO TO 10000                                               16
2000      IF(LCORE .EQ. 0) GO TO 2050                               17
          LSUB = 0                                                  18
          CALL NORMAL                                               19
2050      KD = NOLA(4, LMAXP1)                                      20
          READ(9'KD) NAME(1), NAME(2), ICNAM                        21
```

```
C                                                                   RESULT
          CALL UNRAVL                                               22


C***********************************************************************
C     Step 2 :
C
C          LMAX       : Number of subproblems.
C          MAST       : Indicates the phase of the D-W algorithm being executed.
C                       '2' indicates Phase 3.
C          NROW       : Number of rows in the problem being solved.
C          NROWO      : Number of coupling rows.
C          NCOLO      : Number of coupling columns.  That is, the number of columns in
C                       A1_0 (cf. § 2.5.3).
C          JH(*)      : Indicates which column is basic in the *th row.
C          ICOLS(*)   : Indicates the subproblem that generated the *th proposal.
C          Z(*, *)    : Normally the proposal buffer.  In RESULT it is used as a buffer to
C                        compute the coupling rhs of each subproblem.
C          A(*)       : Data array.
C          X(*)       : Solution array.
C          YA(*, 6)   : Contains dual variable for the coupling rows in D-W Phase 2.
C
C     (a) Set MAST to 2 to indicate that a Phase 3 operation is being done.  [line 24]
C     (b) Initialize the Z(*, *) buffer.  [lines 26 - 27]
C     (c) Consider a row and obtain the column that is basic in this row.  [line 29]
C     (d) Obtain the subproblem origin of a master problem's column.  [line 30]
C     (e) If the column was generated by the subproblem being considered then compute the
C         allocation due to this column for all resources.  [lines 31 - 34]
C         Else go to (c) to take up the next row unless all rows have been considered.
C     (f) Write the allocation on the unit 8 work file.  [line 36]
C     (g) Call UNPACK to obtain the rhs of the master in YA(*, 1).  [line 39]
C     (h) Copy the dual variables of the coupling rows into XYZ(*) and the master's rhs
C          into XRHS(*).  [lines 40 - 43]
C     (i) Subtract the amount of resource consumed by the master's coupling columns A1_0.
C         This is done according to (3.6).  [lines 46 - 56]
C
C***********************************************************************
          IF(LMAX .EQ. 0) GO TO 10000                               23
```

```
C                                                          RESULT
        MAST = 2                                             24
        DO 2500 K = 1, LMAX                                  25
        DO 2100 I = 1, NROW                                  26
2100    Z(I, 1) = 0.                                         27
        DO 2400 L = 1, NROW                                  28
        IV = JH(L)                                           29
        IF(ICOLS(IV) .NE. K) GO TO 2400                      30
        MM = LA(IV + 1) - 1                                  31
        NNN = LA(IV)                                         32
        DO 2300 I=NNN, MM                                    33
2300    Z(IA(I), 1) = Z(IA(I), 1) + A(I)*X(L)                34
2400    CONTINUE                                             35
        WRITE(8) (Z(N, 1), N = 1, NROWO)                     36
2500    CONTINUE                                             37
        REWIND 8                                             38
        CALL UNPACK(RHSCOL, 1)                               39
        DO 2520 I = 1, NROW                                  40
        IF(I .LE. NROWO) ITEMP(I) = - 1                      41
        XYZ(I) = YA(I, 6)                                    42
2520    XRHS(I) = YA(I, 1)                                   43
        ITEMP(IOBJ) = 0                                      44
        DO 2550 I = 1, NROW                                  45
        IV = JH(I)                                           46
        IF(IV .GT. NROWO) GO TO 2530                         47
        IF(X(I) .LE. ZTCAND) GO TO 2530                      48
        ITEMP(IV) = 0                                        49
2530    IF(IV .GT. NCOLO) GO TO 2550                         50
        LL = LA(IV)                                          51
        KK = LA(IV + 1) - 1                                  52
        DO 2540 J = LL, KK                                   53
        IR = IA(J)                                           54
2540    XRHS(IR) = XRHS(IR) - A(J)*X(I)                      55
2550    CONTINUE                                             56
        DO 2560 I = 1, NROWO                                 57
2560    XYZ(I) = YA(I, 6)                                    58
```

```
C************************************************************************
C     Step 3 :                                                  RESULT
C
C          LSUB      : Index of the subproblem being solved.
C          LMAX      : Number of subproblems.
C          NROW      : Number of rows in the subproblem being solved.
C          NROWO     : Number of coupling rows.
C          YA(*, 6)  : Contains the dual vector.  The first NROWO elements of YA(*, 6)
C                      contain the dual vector of the coupling rows.
C          XYZ(*)    : Contains the dual vector of the D-W Phase 2 coupling row in the
C                      first NROWO elements.  The other elements contain the Phase 2
C                      dual vector of the non-coupling rows of the subproblem being
C                      solved.
C          ISTYPE(*) : Indicates the type of the *th constraint. '1' indicates equality.
C          MAST      : Indicates the stage of the algorithm being executed.
C          XRHS(*)   : The amount of each resource left after solving a subproblem.
C
C     Reference will be made to "solving subproblems" part of "description". Note that the
C     reconstruction of only one $x_r^*$ is described below in steps (a) - (e).
C
C     (a) Read the problem and call INVERT to update the solution.  [lines 62 - 66]
C     (b) Now $\pi_r^*$ (cf. step 2.2) is to be reconstructed. For this $\pi_0^k$ is required. However,
C         after INVERT, $\pi_0^k$, which was residing in the first NROWO elements of YA(*, 6)
C         has been destroyed and needs to be copied from XYZ(*).  [lines 68 - 69]
C         Call FORMC and BTRAN to compute $\pi_r^*$.  [lines 70 - 71]
C         Save the dual vector [$\pi_0^*, \pi_r^*$] in XYZ(*).  [lines 72 - 73]
C         Note that since this is not a Phase 3 step, MAST is set to 1.  [line 67]
C     (c) Now step 2.3 is to be executed. Hence, reset MAST to 2.  [line 74]
C         Set type of the constraints $A1_r x_r = y_r$ to equality.  [lines 41, 75 - 76]
C         Call NORMAL to reconstruct $x_r^*$.  [line 78]
C     (d) Update XRHS(*) using (3.6).  [lines 81 - 90]
C     (e) Call UNRAVL to output $x_r^*$.  [line 91]
C     (f) Compute the infinity norm of XRHS(*) and print it.  [lines 98 - 103]
C
C************************************************************************
```

```
C                                                               RESULT
            NTEMP(9) = NT                                          59
            NTEMP(10) = NX                                         60
            DO 3000 I = 1, LMAX                                    61
            LSUB = I                                               62
            LSUP = I                                               63
            CALL CHANGE(LSUP)                                      64
            READ(8) (Z(N, 1), N = 1, NROWO)                        65
            CALL INVERT                                            66
            MAST = 1                                               67
            DO 2570 JK = 1, NROWO                                  68
2570        YA(JK, 6) = XYZ(JK)                                    69
            CALL FORMC                                             70
            CALL BTRAN                                             71
            DO 2580 JK = 1, NROW                                   72
2580        XYZ(JK) = YA(JK, 6)                                    73
            MAST = 2                                               74
            DO 2600 J = 1, NROWO                                   75
            ISTYPE(J) = ITEMP(J)                                   76
2600        CONTINUE                                               77
            CALL NORMAL                                            78
            KD = NOLA(4, LSUB)                                     79
            READ(9'KD) NAME(1), NAME(2), ICNAM                     80
            DO 2800 JK = 1, NROW                                   81
            IV = JH(JK)                                            82
            LL = LA(IV)                                            83
            KK = LA(IV + 1) - 1                                    84
            DO 2700 JRHS = LL, KK                                  85
            IR = IA(JRHS)                                          86
            IF(IR .GT. NROWO) GO TO 2700                           87
            XRHS(IR) = XRHS(IR) - A(JRHS)*X(JK)                    88
2700        CONTINUE                                               89
2800        YA(JK, 6) = XYZ(JK)                                    90
            CALL UNRAVL                                            91
            NTEMP(9) = NTEMP(9) + NT                               92
            NTEMP(10) = NTEMP(10) + NX                             93
3000        CONTINUE                                               94
```

C		RESULT
	LSUP = 0	95
	CALL CHANGE(LSUP)	96
	ERMAX = 0.0	97
	DO 4000 I = 1, NROWO	98
	IF(ISTYPE(I) .EQ. 0) GO TO 4000	99
	IF(DABS(XRHS(I)) .GT. ERMAX) ERMAX = DABS(XRHS(I))	100
4000	CONTINUE	101
	WRITE(6, 4500) ERMAX	102
4500	FORMAT(' MAXIMUM ROW ERROR IN COMMON ROWS = ', E14.5)	103
	TIMER1 = NTEMP(9) / 1000.	104
	TIMER2 = NTEMP(10) / 1000.	105
	WRITE(6, 5000)TIMER1	106
5000	FORMAT(' TOTAL TRANSFORMATION TIME = ', F8.2, ' SECONDS')	107
	WRITE(6, 6000)TIMER2	108
6000	FORMAT(' TOTAL INPUT/OUTPUT TIME = ', F8.2, ' SECONDS')	109
10000	RETURN	110
	END	111

3.18.   Subroutine UNPACK

3.18.1. Major Task

Retrieves a column stored in packed form in the data array A(*).

3.18.2. Inputs

1. KVEC, giving the column position in the buffer YA(*, *) into which the data column is to be unpacked. Note that throughout this subroutine the term "data column" will be used to indicate the column in the array A(*) that is to be unpacked.

2. IV, giving the index of the data column that is to be unpacked.

3.18.3. Called By

NORMAL, INVERT, UNRAVL.

3.18.4. Comments

158

```
C********************************************************************
C      YA(*, *) : Buffer into which the data column is to be unpacked.
C      KVEC    : Unpacks the data column into the KVEC$^{th}$ column of YA(*, *).
C      A(*)     : Data array.
C      IA(*)    : Indicates the row index of the *$^{th}$ element of A(*).
C      LA(*)    : Indicates the position in A(*) where the first non-zero element of the *$^{th}$ data
C                 column resides.
C
C      1. Store the zero vector in YA(*, KVEC).  [lines 1 - 3]
C      2. Set LL to the starting position in A(*) of the data column and set KK to the ending
C         position in A(*) of the data column.  [lines 4 - 5]
C      3. If there are no elements in the data column then RETURN.  Else proceed.  [line 6]
C      4. For each non-zero element A(i) in the data column do :
C              - Set IR to IA(i).  Assign A(i) to YA(IR, KVEC). [lines 8-9]
C         End For
C
C********************************************************************
C                                                                    UNPACK
               DO 100 I = 1, NROW                                     1
               YA(I, KVEC) = 0.                                       2
  100          CONTINUE                                               3
               LL = LA(IV)                                            4
               KK = LA(IV + 1) - 1                                    5
               IF(KK .LT. LL) GO TO 500                               6
               DO 200 I = LL, KK                                      7
               IR = IA(I)                                             8
               YA(IR, KVEC) = A(I)                                    9
  200          CONTINUE                                               10
  500          RETURN                                                 11
               END                                                    12
```

3.19. Subroutine UNRAVL

3.19.1. Major Task

Outputs the primal solution vector $\mathbf{x}_r$ ($r = 0,..., R$) corresponding to the rth problem.

3.19.2. Inputs

1. The index of the subproblem.

2. The solution vector x_r in the array X(*).

3. The dual vector in the buffer YA(*, 6).

3.19.3. Outputs

1. The row names.

2. The column names or in case the column is a proposal the subproblem origin of the proposal.

3. The dual vector, the rhs vector and the primal solution vector.

3.19.4. Called By

RESULT.

3.19.5. Calling

UNPACK to unpack the rhs vector.

3.19.6. Description

The solution vector and the dual vector are provided by subroutine RESULT. The rhs vector is obtained by calling UNPACK and in case of the master problem, '1's are placed in positions corresponding to convexity constraints. The output for each problem (master or subproblem) is in a compact form listing on each line a basic variable, its solution value, the row it is basic in, the dual variable and the rhs for that row.

All row and column names are stored in two parts. In a subproblem, all names are of type character. Their output format is straight forward. For the master problem, parts of certain names (e.g. those for proposals) are generated by the program as integers. Other parts are of type character. The structure of the master problem is illustrated in Figure 3.17 where NCOLO is the number of coupling columns, NROWO is the number of coupling rows and NROW is the number of rows (including convexity) in the master problem. The various composition of names are summarized in Table 3.1.

From Table 3.1 we see that there are two kinds of row names (CC and CI) and three kinds of column names (CC, CI and I I). Therefore, depending on what kind of column is basic in each row, a total of six cases are possible. A separate format statement is provided in each case. The cases and line numbers in the code for them are summarized in Table 3.2

Name for	Type of Part 1	Type of Part 2	Type Abbreviation
Coupling Rows	Character	Character	CC
Convexity Rows	Character	Integer	C I
Coupling Logicals	Character	Character	CC
Convexity Logicals	Character	Integer	C I
Coupling Columns	Character	Character	CC
Proposals Columns	Integer	Integer	I I

Table 3.1 Composition of names in the master problem

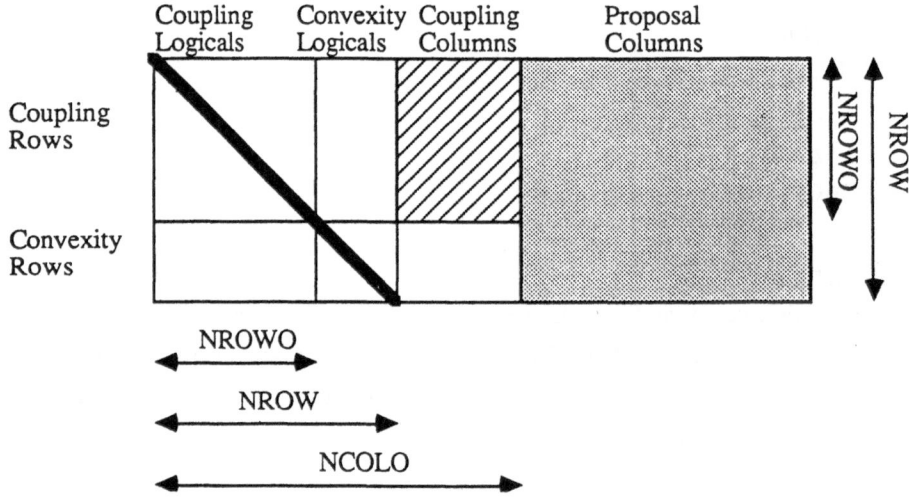

Figure 3.17   Structure of master problem

Kind of Row	Kind of Basic Column	Lines
CC	CC	[23 - 25]
CC	C I	[28 - 29]
C I	C I	[32 - 34]
C I	CC	[35 - 37]
CC	I I	[39 - 41]
C I	I I	[42 - 44]

Table 3.2 Six combinations of row and column names in the master problem

3.19.7. Comments

```
C*********************************************************************
C      NROW  : Number of rows in the problem being considered.
C      NROWO: Number of coupling rows.
C      NCOLO : Number of coupling columns.
C      YA(*, 6) : Contains the dual vector.
C      YA(*, 1) : Contains the rhs vector.
C      X(*)     : Contains the output solution vector.
C      JH(*)    : Indicates the column that is basic in the *th row.
C      IV       : Index of the column that is basic in the row being considered.
C      LMAX     : Number of subproblems.
C      LSUB     : Index of the problem being considered.  '0' indicates master problem.
C      RHSCOL: The column that is used as the rhs of the problem.
C
C      1. Call UNPACK to unpack the rhs.  If LSUB indicates the master problem then
C         store '1's in positions corresponding to convexity constraints.  If LSUB indicates
C         a subproblem then copy the allocation rhs y_r. [lines 6, 9 - 10, 13 - 14]
C      2. For each row, the name of the column basic in that row, the value of the basic
C         variable, the name of the row, the value of its dual variable, and its right-hand side
C         coefficient are printed. The format used is according to Table 3.2
C*********************************************************************
C                                                                    UNRAVL
            EQUIVALENCE (ICNAM(1), A(5001))                          1
            WRITE(6, 8000) (NAME(I), I = 1, 2)                       2
8000        FORMAT(1H1, 'SUBPROBLEM ', 2A4 //)                      3
            WRITE(6, 8100)                                          4
8100        FORMAT(6H JH(I), 16X, 5HVALUE, 9X, 9HROW             5
           1 NAMES, 9X, 5HPI(I), 12X, 3HRHS )
            CALL UNPACK(RHSCOL, 1)                                  6
            IF(LMAX .EQ. 0 .OR. MAST .EQ. 0) GO TO 500              7
            IF(LSUB .GT. 0) GO TO 200                               8
            DO 100 I = 1, LMAX                                      9
100         YA((NROWO + I), 1) = 1.                                 10
            GO TO 500                                               11
200         IF(MAST .NE. 2) GO TO 500                               12
            DO 300 I = 1, NROWO                                     13
300         YA(I, 1) = Z(I, 1)                                      14
```

```
C                                                                  UNRAVL
500          DO 1000 I = 1, NROW                                      15
             IV = JH(I)                                               16
             IRNAM1 = ICNAM(I, 1)                                     17
             IRNAM2 = ICNAM(I, 2)                                     18
             IF(LSUB .EQ. 0 .AND. IV .GT. NCOLO) GO TO 800            19
             ICNAM1 = ICNAM(IV, 1)                                    20
             ICNAM2 = ICNAM(IV, 2)                                    21
             IF(LSUB .EQ. 0 .AND. MAST .NE. 0) GO TO 600              22
550          WRITE(6, 8200) ICNAM1, ICNAM2, X(I), IRNAM1, IRNAM2,     23
      1 YA(I, 6), YA(I, 1)
8200         FORMAT(1H , A4, A4, 8X, G16.8, 4X, A4, A4, 4X, 2G16.8 )   24
             GO TO 1000                                               25
600          IF(I .GT. NROWO ) GO TO 700                              26
             IF(IV .LE.NROWO .OR. IV .GT. NROW) GO TO 550             27
             WRITE(6, 8300) ICNAM1, ICNAM2, X(I), IRNAM1, IRNAM2,     28
      1 YA(I, 6), YA(I, 1)
8300         FORMAT(1H , A4, I4, 8X, G16.8, 4X, A4, A4, 4X, 2G16.8 )   29
             GO TO 1000                                               30
700          IF(IV .LE. NROWO .OR. IV .GT. NROW) GO TO 750            31
             WRITE(6, 8400) ICNAM1, ICNAM2, X(I), IRNAM1, IRNAM2,     32
      1 YA(I, 6), YA(I, 1)
8400         FORMAT(1H , A4, I4, 8X, G16.8, 4X, A4, I4, 4X, 2G16.8 )   33
             GO TO 1000                                               34
750          WRITE(6, 8500) ICNAM1, ICNAM2, X(I), IRNAM1, IRNAM2,     35
      1 YA(I, 6), YA(I, 1)
8500         FORMAT(1H , A4, A4, 8X, G16.8, 4X, A4, I4, 4X, 2G16.8 )   36
             GO TO 1000                                               37
800          IF(I .GT. NROWO) GO TO 850                               38
             WRITE(6, 8600) IV, ICOLS(IV), X(I), IRNAM1, IRNAM2,      39
      1 YA(I, 6), YA(I, 1)
8600         FORMAT(1H , 2I4, 8X, G16.8, 4X, A4, A4, 4X, 2G16.8 )      40
             GO TO 1000                                               41
850          WRITE(6,8700) IV, ICOLS(IV), X(I), IRNAM1, IRNAM2,       42
      1 YA(I, 6), YA(I, 1)
8700         FORMAT(1H , 2I4, 8X, G16.8, 4X, A4, I4, 4X, 2G16.8 )      43
1000         CONTINUE                                                 44
```

		UNRAVL
C		
	TIMER = NT/1000.	45
	WRITE(6, 8800)TIMER	46
8800	FORMAT (4H EOF, /, '0TRANSF TIME =' , F8.2, 8H SECONDS)	47
	TIMER = NX/1000.	48
	WRITE(6, 9000)TIMER	49
9000	FORMAT(' INPUT/OUTPUT TIME = ', F8.2, ' SECONDS'/'1')	50
	RETURN	51
	END	52

3.20. Subroutine UPBETA

3.20.1. Major Task
Updates the solution after a simplex basis change.

3.20.2. Input
Index of the entering column.

3.20.3. Called By
NORMAL.

3.20.4. Comments

```
C********************************************************************
C     YA(*, *)   : Buffer containing the entering column..
C     KVEC       : The column index of the buffer YA(*, *) that contains the entering
C                   column.
C     IVOUT      : Index of the leaving column.
C     KINBAS(*) : If the *th column is basic it indicates the row in which it is basic.
C     X(*)       : Contains the solution vector.
C     DP         : Gives the allowable increase in the entering variable.
C     IROWP      : Index of the pivot row.
C     DE         : Value of the IROWPth element of the current rhs.
C     DCMIN(*)  : Reduced cost of the *th column of the buffer YA(*, *)
C     DOB        : Change in the objective value.
C
```

```
C      1. Compute DP.  [lines 1 - 3]
C      2. The value of the new basic variable for the pivot row IROWP should be the allowable
C           increase in the entering variable.  Hence set X(IROWP) to DP.  [line 4]
C      3. For each non-pivot row do
C                - Set X(i) <-- X(i) - YA(i, KVEC)*X(IROWP).  If |X(i)| is less
C                     than the zero tolerance then set it to zero.  [lines 7 - 9]
C           End For
C      4. Compute the increase in the objective due to this pivot change provided the problem
C           being solved is the subproblem.  Recall that this would determine the reduced cost of
C           the proposal if the current solution is sent as a proposal.  [lines 11 - 13]
C
C*********************************************************************
C                                                              UPBETA
         IROWP = KINBAS(IVOUT)                                    1
         DE = X(IROWP)                                            2
         DP = DE/YA(IROWP, KVEC)                                  3
         X(IROWP) = DP                                            4
         DO 1000 I = 1, NROW                                      5
         IF (I .EQ. IROWP) GO TO 1000                            6
         DE = X(I)                                               7
         X(I) = DE - YA(I, KVEC)*DP                             8
         IF (DABS(X(I)) .LE. 1.0D-15) X(I) = 0.0D0              9
1000     CONTINUE                                               10
         IF(KFASE .EQ. 1 .OR. LSUB .EQ. 0) GO TO 2000          11
         IF (MAST .EQ. 0) GO TO 2000                            12
         DOB = DOB + DCMIN(KVEC)*DP                             13
2000     RETURN                                                 14
         END                                                    15
```

3.21. Subroutine VECTOR

3.21.1. Major Task

Reads a specified problem into core from DAD or writes a problem in core to DAD.

3.21.2. Input

The parameter K1 which indicates a write if its value is one and a read if its value is two.

3.21.4. Called By

CHANGE, INDATA, NORMAL.

3.21.5. Comments

```
C*********************************************************************
C     K1       : Parameter.  '1' indicates write and '2' indicates read.
C     AX(*)    : Equivalent to A(*).
C     IAX(*)   : Equivalent to IA(*).
C     NBX(*)   : Equivalent to NB(*).
C*********************************************************************
C                                                              VECTOR
      GO TO (100, 200) K1                                          1
100   WRITE (9'KD) AX, IAX, NBX                                    2
      GO TO 1000                                                  3
200   READ (9'KD) AX, IAX, NBX                                    4
1000  RETURN                                                      5
      END                                                         6
```

3.22. Subroutine WRETA

3.22.1. Major Task

Writes an eta to the data array.

3.22.2. Input

KVEC, for the column position in YA(*, *) containing the eta vector.

3.22.3. Called By

NORMAL.

3.22.4. Description

WRETA writes in the array A(*) the eta vector for the change of basis. Suppose the pth column of the matrix **A** is the entering column. Let its update with respect to the current basis be $[a_{1p}, a_{2p},..., a_{pp},..., a_{np}]^t$. Then the eta vector generated due to this column is :

$$[-a_{1p}/a_{pp},..., 1/a_{pp},..., -a_{np}/a_{pp}]^t \qquad (3.42)$$

The pivot element of the eta vector $1/a_{pp}$ is stored first. Its row index specifes the pivot row. This saves the need for storing the pivot row index separately. Then, each of the remaining non-zero element a_{ip} (and *not* $-a_{ip}/a_{pp}$) is stored. The necessary arithmatic is efficiently built into the subroutines that use the etas (BTRAN and FTRAN).

3.22.5. *Comments*

```
C*********************************************************************
C     A(*)            : Array containing the data and eta elements.
C     IA(*)           : Indicates the row index in the matrix A of the *th element of the
C                         array A(*).
C     NELEM           : Number of non-zero elements in the array A(*).
C     YA(*, KVEC)     : Contains the updated pivot column a_p.
C     IROWP           : Index of the pivot row.
C     NETA            : Numbre of etas.
C
C     1. Incorporate the eta element 1/a_pp into the array A(*).  [lines 2 - 4]
C     2. Incorporate each non-zero element a_ip (for all i ≠ p) into the array  A(*). [lines 6 - 10]
C     3. Increment NETA, the number of etas.  [line 12]
C     4. Mark the ending position of this eta.  [line 13]
C
C*********************************************************************
C                                                               WRETA
            DATA ZTETA/1.D-15/                                     1
            NELEM = NELEM + 1                                      2
            IA(NELEM) = IROWP                                      3
            A(NELEM) = 1. / YA(IROWP, KVEC)                        4
            DO 1000 I = 1, NROW                                    5
            IF (I .EQ. IROWP) GO TO 1000                           6
            IF (DABS(YA(I, KVEC)) .LE. ZTETA) GO TO 1000           7
            NELEM = NELEM + 1                                      8
            IA(NELEM) = I                                          9
            A(NELEM) = YA(I, KVEC)                                10
```

```
C                                                              WRETA
1000      CONTINUE                                               11
          NETA = NETA + 1                                        12
          LE(NETA + 1) = NELEM + 1                               13
          RETURN                                                 14
          END                                                    15
```

CHAPTER 4

Portability Issues

In this chapter issues concerning portability of the code are addressed. Since the code is written in FORTRAN IV it is easily portable. However, some parts of the code are machine dependent and would require modification when installed on different machines. The following paragraphs identify these parts.

4.1 Direct Access Device

The ultimate advantage of decomposition is to solve large problems with bulky data sets that cannot be accommodated in core memory simultaneously. Hence, direct access I/O becomes a necessary part of any robust implementation. Since direct access I/O is machine dependent it is not possible to have a universally portable code.

The statement defining the direct access device (DAD) normally contains a field for specifying the record length. On the IBM/370 series machines this field is specified in terms of bytes. But on the VAX-VMS the record length must be in terms of four-byte words and on a UNIX-based machine it has be in terms of bits. Also, on the IBM/370, programming direct access I/O is easy since any message can be written (or read) simply by specifying the starting record number in the statement:

WRITE(Unit number ' Starting record number) "Message."

However, on the VAX-VMS and on the UNIX based systems this would cause an error whenever the size of the message exceeds the record length.

In DECOMP, these machine-dependent differences require major changes in the subroutine VECTOR. The pseudo-code to perform some of the functions of VECTOR on a VAX-VMS or similar system is as follows.

KD <-- Starting record number.
RECL <-- Record length in bytes.
DATSIZ <-- Size of the message in bytes.
LAST <-- Last component of A(*) that is to be written

A(*) <-- Array containing the message.

C

C It is assumed for simplicity that each component of A(*) is a byte long.

C KD gets incremented after the execution of a write.

C

```
J <-- 0
DO WHILE (DATSIZ > RECL)
            WRITE(Unit Number'KD) A(J+1:J+RECL)
            DATSIZ <-- DATSIZ - RECL
            J <-- J + RECL
END
WRITE (Unit Number'KD) A(J+1:LAST)
```

Note that the actual code will have to be somewhat more complicated because the psuedo-code given above works correct only if all the elements to be written to DAD are contiguous components of the array A(*). This is not true for the master problem since data elements are followed by a gap to accommodate the expansion of the master due to column generation. See Figure 4.1.

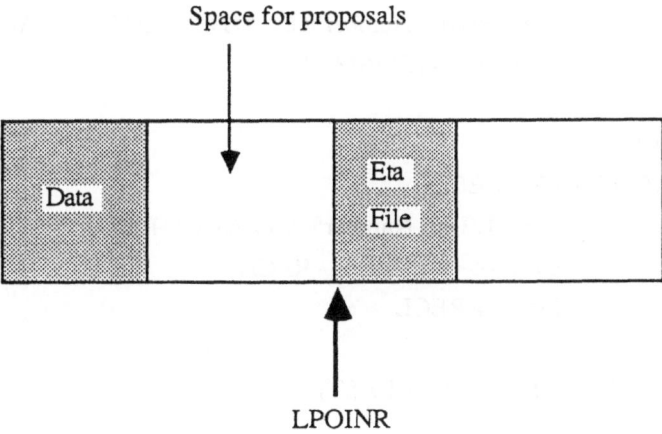

Figure 4.1 Organization of the data array for the master problem

Therefore, the pseudo-code to perform the function of VECTOR in the case of the master problem has to be modified as follows.

```
        KD <-- Starting record number.
        RECL <-- Record length in bytes.
        DATSIZ <-- Size of the message (not including etas).
        ETASIZ <-- Size of the eta file.
        A(*) <-- Array containing the message to be written.
        LN <-- Indicates where in A(*) the eta file begins.
        LAST <-- Last element of A(*) that is to be written.
        J <-- 0
C
C  For simplicity each component of A(*) is assumed to be a byte long.
C  KD gets incremented every time a WRITE is executed.
C
        J <--- 0
        DO WHILE (DATSIZ ≥ RECL)
                        WRITE(Unit Number'KD) A(J+1:J+RECL)
                        DATSIZ <-- DATSIZ - RECL
                        J <-- J + RECL
        END
        IF (DATSIZ ≠ 0) THEN
                        L <-- RECL - DATSIZ
                        WRITE(Unit Number'KD) A(J+1:J+DATSIZ), A(LN+1:LN+L)
                        ETASIZ <-- ETASIZ - L
        END
        J <-- LN + L
        DO WHILE (ETASIZ > RECL)
                        WRITE(Unit Number'KD) A(J+1:J+RECL)
                        ETASIZ <-- ETASIZ - RECL
                        J <-- J + RECL
        END
        WRITE(Unit Number'KD) A(J+1:LAST)
```

4.2   NAMELIST Statements

Another modification that may be necessary involves the input format for NAMELIST statements. For example, on the VAX-VMS the statements :

```
NAMELIST/NM/ X, Y
READ(Unit Number, *) NM
```

should be replaced by :

```
NM
        X = Value of X
        Y = Value of Y
$END
```

On a UNIX-based machine the values of X and Y need to separated by blank spaces.

CHAPTER 5

User's Guide

DECOMP is a FORTRAN IV code of the Dantzig-Wolfe algorithm for solving linear programming problems of the form

$$\text{Minimize} \quad z = \sum (c_r x_r)$$

subject to

$$\sum (A1_r x_r) \quad = b_0$$
$$A2_r x_r \quad = b_r \quad ; \quad r = 1,..., R$$
$$x_r \quad \geq 0 \quad ; \quad r = 1,..., R$$

where c_r is 1 by n_r, b_r is m_r by 1 and all other vectors and matrices are of compatible dimensions. DECOMP was first coded by C. Winkler based on J.A. Tomlin's LPM1 linear programming code at the Systems Optimization Laboratory (SOL) at Stanford University. It has been further developed by J.K. Ho and E. Loute at the Center for Operations Research and Econometrics (CORE), University of Louvain, Belgium.

5.1. Dimension of Arrays

The arrays within the program are dimensioned to accommodate from zero up to ten subproblems. Each subproblem can have up to 400 rows (including the m_0 coupling rows and the objective row), 1000 columns (including logicals, structurals, and rhs columns) and up to 10,000 non-zero elements in the data array (including subproblem data and eta file). These restrictions also apply to master problem with the following remarks :

1. The master problem can have up to 99 rows (including the objective and coupling rows) besides the convexity rows.

2. The number of columns (including logicals, proposals and rhs columns) must be less than 1000.

DECOMP uses three external files. A direct access file (unit 9) of 106 records with

13028 bytes/record is used to store the master problem and subproblem data on disk. A sequential file (unit 8) is needed for proposal buffering and in Phase 3 of the D-W algorithm. The data of the whole problem is read from a unit 2 sequential file. A typical set of JCL instructions to run DECOMP on an IBM mainframe under the MVS operating system is given in § 5.4.2.

DECOMP is written in FORTRAN IV with some IBM FORTRAN features (eg. DEFINE FILE etc...). It uses a timer routine TICTAC with an INTEGER*4 output argument. The difference in the arguments corresponding to two calls gives the CPU time spent between the two calls. If a timer is not available the user can replace it by the following dummy routine:

```
SUBROUTINE TICTAC(ITIME)
INTEGER*4 ITIME
ITIME = 0
RETURN
END
```

5.2. Input Data

The input data is read from the sequential unit 2 file and consists of two sections : (a) the control parameters and (b) the master and subproblem data. The control parameters are variables on the NAMELIST statement called PARAM and are listed below.

LMAX   : the number of subproblems. It *must* be specified.

LPOINR   : pointer to the start of the eta file in the master problem data array (default is 5000).

IOBJ   : index of the objective row to be used (default is 1).

USERHS   : index of the rhs column to be used from the set of user-supplied rhs columns (default is 1).

KINP   : the unit number of the input device (default is 2).

KSTR   : a strategy parameter for D-W algorithm (default is 2).

           When KSTR = 1, each subproblem sends one proposal at optimality and two proposals at unboundedness.

           When KSTR = 2, each subproblem is solved until MAXCA proposals have been sent to the master, unless optimality has been reached before. With this strategy a major cycle is interrupted once a total of MAXNCA proposals have been reached. The master problem is then solved and another major cycle is started with the next subproblem.

When KSTR = 3, then in the first major cycle all subproblems are solved until they have provided up to MAXCA proposals and together no more than MAXNCA proposals and in subsequent major cycles only one subproblem is solved per cycle in a cyclic manner.

When KSTR = 4, then in a major cycle a subproblem is solved allowing for up to MAXCA proposals; if the subproblem is not optimized then a cycle is interrupted and a new one is started with that subproblem as the first subproblem.

When KSTR = 5, the strategy is similar to KSTR = 2 except that subproblems are solved to optimality with only the last MAXCA proposals sent to the master.

MAXNCA	: maximum number of proposals that can be sent to the master in a major cycle (Limit is 150 and default is 100).
MAXCA	: maximum number of proposals that can be sent to the master by a particular subproblem during a major cycle (default is 10).
NCANFR	: frequency of proposal generation (default is 15).
INVFRQ	: frequency of basis re-inversion in the Simplex routine (default is 30).
INVFMA	: frequency of basis re-inversion for the master problem (default is 20).
KMULT	: the number of columns that can be chosen for multiple pricing (maximum is 5 and default is 3).
ITOL	: switch to indicate whether tolerances are to be read.
IRST	: restart paramater indicating when non-zero that the decomposition algorithm is to be restarted at the cycle when it was saved in a previous run (default is 0).
ISAVE	: parameter which indicates when non-zero that restart information is to be saved when the algorithm is stopped (default is 1).
ISOL	: when set to one it indicates that a Phase 3 is to be performed (default is 0).
ICLAST	: maximum number of cycles to be performed (default is 999)
NPH	: when set to '1' it indicates that each subproblem is solved to optimality and sends one proposal to the master before the first major cycle (default is 1).
RPERC	: tolerance used in the proposal generation machanism (cf. Variable Dictionary) (default is 40 %).
RSTOP	: tolerance used in the primal-dual stopping criterion (default is .01 %)

The following tolerances can be specified if ITOL ≠ 0 via a NAMELIST called TOL : ZTOLZE, ZTOLPV, ZTCOST, ZTOLRJ, ZTCAND. The meaning and default values of these tolerances are given in the variable dictionary.

The master problem data set follows the specification of the control parameters. This is followed by the subproblem data set. All of these (master and subproblem data set) are in standard MPS format (cf. IBM MPSX/370 Basic Reference Manual, SH19-1127-0). Specific requirements for each data set are given below (cf. § 5.4 for an example).

Case 1 : (Master problem)

1. The coupling rows are to be declared in the master problem's ROWS section.

2. Only the elements corresponding to $A1_0$ (coupling columns) are to be specified in the COLUMNS section. If there are no coupling columns then the COLUMNS section would just be the keyword "COLUMNS"

3. If several rhs columns are defined in the problem, they must be declared in the master and in each subproblem in the same order.

4. Even if a rhs column has no nonzeros in the master it *must* be declared in the rhs section of the master by entering any zero element.

5. The rhs data corresponding to coupling rows must appear only in the rhs section of the master problem.

Case 2 : (Subproblem)

1. Declaration of the coupling rows in the subproblem data is optional.

2. The COLUMNS section of the rth (r = 1,..., R) subproblem consists of elements in $[A1_r^t, A2_r^t]^t$.

3. If several rhs columns are defined in the problem, they must be declared in the master and in each of the subproblems in the same order.

4. Even if a rhs column has no nonzeros in a subproblem it *must* be declared in the rhs section of the subproblem by entering any zero element.

5.3. Output Data

The output consists of various statistics for each step of the D-W algorithm. The different kinds of statistics are explained first. Then the output for individual steps are described. Note that if a timer is not installed the times printed out will always be zero.

(a) Input statistics : problem name, MPS section headers (eg. ENDATA); number of rows; number of structural columns; number of nonzero elements; nonzero density and input time.

(b) Invert statistics : number of eta vectors (etas) and eta elements before and after inversion; the maximum row error; number of nonzeros and structural columns in the basis, number of vectors above and below bump (cf. Subroutine INVERT in § 3.10); and the number of nonzeros and etas in the L and U factors of the basis.

(c) Beginning of problem statistics : problem designation; number of candidates, number of elements in the candidates and mean density of candidates in the case of the master problem.

(d) Iteration statistics : At every simplex iteration until optimality the following is printed.

ITCOUNT	=	iteration number. Followed by 'M' if it is a major iteration in which a full pricing of the columns is performed.
STATUS	=	status of the current basic solution:
		'I' for infeasible;
		'F' for feasible;
		'N' if no feasible solution exists;
		'U' if unbounded;
		' ' (blank) if optimal.
OBJ. VALUE	=	current value of the objective function.
VECIN	=	index of the column entering the basis.
FROM	=	used in master problem to indicate the subproblem origin of a proposal column.
VECOUT	=	index of the column leaving the basis.
DJ	=	reduced cost of the vector entering the basis.
NETA	=	number of etas in the eta file.
NELEM	=	number of nonzeros in the eta file.
TIME	=	CPU time.
NO. CAND	=	number of proposals generated by the subproblem in the current cycle.
CRIT	=	criterion under which the proposal was generated (cf. § 3.3.6):
		' ' = no proposal;
		'R' = percentage of improvement;
		'F' = frequency test;

'A' = Test A;

'B' = Test B;

'O' = optimality;

'U' = unboundedness.

(e) End of problem statistics : lower bound and relative difference (only for the master problem).

(f) Beginning of cycle statistics : cycle number; time for the previous cycle and the time in the subroutine MASTER.

(g) End of D-W Phase 2 statistics : the number of proposals generated under each criterion (cf. IPROP in Variable Dictionary and Subroutine CHECK).

(h) Solution statistics for master problem :

JH(*) = name of the basic column in the $*^{th}$ row; or the column index and subproblem origin of a basic proposal column.

VALUE = value of the basic variable.

ROWNAME = name of the row.

PI(*) = value of the dual variable for the $*^{th}$ row.

RHS = right-hand side coefficient for the current row.

Output during Step 1 of D-W Phase 1:

    - Program control parameters;

    - For each problem i including the master:

        - Input statistics;

        - Invert statistics and iteration statistics if problem i is a subproblem;

    End For;

    - Total input time for the entire problem.

Output during Step 2 of D-W Phase 1:

    The output would be the same as that of D-W Phase 2 except that the relative difference and the lower bounds are not printed.

Output during D-W Phase 2:

    - For each cycle:

- Beginning of cycle statistics;
- For each problem beginning with the master problem:
  - Invert statistics if subroutine INVERT is called;
  - Beginning of problem statistics;
  - Iteration statistics;
  - End of problem statistics;
End For;
End For.

Output during D-W Phase 3:

The following is printed if the parameter ISOL is set to one.
- For each subproblem:
  - Problem name;
  - Invert statistics;
  - Beginning of problem statistics and iteration statistics;
  - Solution statistics for subproblem (same format as (h) above);
  - Transformation time, input and output time;
End For;
- Maximum row error in common rows (cf. Subroutine RESULT).

5.4. An Example

Consider the problem :

Minimize	$-X0$	$-X1$	$-X2$	$-2Y1$	$-Y2$		
subject to							
	$+X0$	$+X1$	$+2X2$	$+2Y1$	$+Y2$	$\leq$	40
		$+X1$	$+3X2$			$\leq$	30
		$2X1$	$+X2$			$\leq$	20
				$Y1$		$\leq$	10
				$Y1$	$+Y2$	$\leq$	10
					$Y2$	$\leq$	15
			X0, X1, X2, Y1, Y2			$\geq$	0

Suppose that the seven rows (starting with the objective) are named OBJ, RM1, RS11, RS12, RS21, RS22 and RS23 respectively and the five columns are named X0, X1, X2, Y1 and Y2 respectively, then the input, the output and a typical set of JCL instructions to run the program on an IBM mainframe under the MVS operating system are given in the following sections.

5.4.1. Input for the Example

For the example above DECOMP requires the following input. All variables in the NAMELIST called PARAM, except LMAX (the number of subproblems), can be left out to take the default values defined in the program. The input for NAMELIST must be formatted to begin with a blank, followed by an ampersand (&) immediately followed by the NAMELIST *name*. Following the name the input data for the NAMELIST variables is coded. Each input data item, separated from one another by a comma, is preceded by a blank and consists of a constant assigned to its corresponding variable. The variable names must appear on these assignments although they can be specified in any order. The end of the NAMELIST input is signalled by coding the keyword "&END" preceded by a blank. If all of the NAMELIST input cannot be accommodated on a single line then the input can be specified in multiple lines. In this case each line must begin with a blank. For further detail, refer to the Programming Guide for VS FORTRAN Version 2 (SC26-4222-3). The input data for each subproblem must conform to standard MPS format.

```
        &PARAM LMAX=2, ISOL=1, KSTR=2 &END
NAME        MASTER
ROWS
   N  OBJ
   L  RM1
COLUMNS
        X0          OBJ         -1.0
        X0          RM1         +1.0
RHS
        RHS1        RM1         +40.0
ENDATA
NAME        SUB1
ROWS
   N  OBJ
   L  RS11
   L  RS12
COLUMNS
        X1          OBJ         -1.0
        X1          RM1         +1.0
        X1          RS11        +1.0
        X1          RS12        +2.0
        X2          OBJ         -1.0
        X2          RM1         +2.0
        X2          RS11        +3.0
        X2          RS12        +1.0
RHS
        RHS1        RS11        +30.0
        RHS1        RS12        +20.0
```

```
ENDATA
NAME          SUB2
ROWS
   N   OBJ
   L   RS21
   L   RS22
   L   RS23
COLUMNS
       Y1        OBJ        -2.0
       Y1        RM1        +2.0
       Y1        RS21       +1.0
       Y1        RS23       +1.0
       Y2        OBJ        -1.0
       Y2        RM1        +1.0
       Y2        RS22       +1.0
       Y2        RS23       +1.0
RHS
       RHS1      RS21      +10.0
       RHS1      RS22      +10.0
       RHS1      RS23      +15.0
ENDATA
```

5.4.2. Job Cards for the Example

A typical set of Job Control Language (JCL) cards for executing DECOMP on an IBM/370 in the MVS environment at the University of Tennessee, Knoxville is shown below. All system dependent items are superscripted with the symbol "¥". If an entire statement is system dependent then the symbol "¥" will appear at the JCL statement number. In statement 6 the value for the parameter PGM stands for compiling and loading the source code using the resident VS FORTRAN compiler. The option LANGLVL=66 in the same statement provides upward compatibility for the FORTRAN IV parts of the code. The JCL card for the direct access file (statement 8) assumes that the unit 9 file is to be discarded immediately after a run of the program. However, if the user wishes to use the restart facility of the code (cf. IRST in § 5.2) then the parameters DSN, DISP, UNIT must be modified appropriately.

```
1¥    //job card
2¥    /*ROUTE PRINT UTKVX.SUNDAR1
3¥    /*JOBPARM LINES=10
4¥    /*UTPARM NONEWS
5¥    /*NOTIFY UTKVX.SUNDAR1
6     //STEP1 EXEC PGM=FORTVCLD¥,
      //  GOREG¥=1024K, LANGLVL=66, FORTOPT=NOSOURCE
7     //FORT.SYSIN DD   *
            Fortran source code (DECOMP)
8     //GO.FT09F001 DD  DSN=&&DAD, SPACE=(13028, 106),
  //  DCB=(DSORG=DA, LRECL=13028, BLKSIZE=13028),
  //  DISP=(NEW, PASS), UNIT=SYSDA
9     //GO.FT08F001 DD
      //  DSN=&&TEMP, UNIT=SYSDA, SPACE=(1612, (100, 10)),
  //  DCB=(RECFM=VBS, BLKSIZE=1612), DISP=(NEW, PASS)
10    //GO.FT06F001 DD
      //  SYSOUT=A, DEST¥=(UTKVX, SUNDAR1)¥, FREE¥=CLOSE¥
11    //GO.FT02F001 DD *
            Input data (as in § 5.4.1)
```

5.4.3. Output for the Example

The print-out for the example is listed below. Note that for the simplex iteration statistics, the lines are wrapped around due to narrower margins than the original computer print-out. Also, comments are overlaid in regular font type and enclosed in square brackets to explain various items in the output. The CPU times in seconds are insignificant for this trivial example and are shown as 0.00 throughout.

```
PROGRAM PARAMETERS

LMAX    2
```

```
LPOINR    5000

IOBJ      1

USERHS    1

KINP      2

KSTR      2

MAXNCA    100

MAXCA     10

NCANFR    15

INVFRQ    30

INVFMA    20

NPH       1

KMULT     3

ITOL      0

ICPI      0

IRST      0

ICLAST    999

ISAVE     1

ISOL      1

NTRAC1    9

NTRAC2    1

ITRK      31

RPERC     40.0 PERCENT

RSTOP     0.100E-01 PERCENT

ZTOLZE    0.100E-06

ZTOLPV    0.100E-05

ZTCOST    0.100E-03

ZTOLRJ    0.100E-04

ZTCAND    0.100E-01 PERCENT
```

```
NAME       MASTER                          [Reading in the master problem]
   ROWS

   COLUMNS

   RHS

   ENDATA


PROBLEM STATISTICS                         [Input summary for master problem]
     4 ROWS

     1 STRUCTURAL COLUMNS

     2 NON-ZERO ELEMENTS
DENSITY =  50.00 PERCENT
INPUT TIME SUBPROBLEM MASTER   =   0.00 SECONDS


NAME            SUB1                        [Reading in Subproblem 1]
   ROWS

   COLUMNS

   RHS

   ENDATA


PROBLEM STATISTICS                         [Input summary for Subproblem 1]
     4 ROWS

     2 STRUCTURAL COLUMNS

     8 NON-ZERO ELEMENTS
DENSITY = 100.00 PERCENT
INPUT TIME SUBPROBLEM SUB1     =   0.00 SECONDS


INVERSION SUBPROBLEM    1            OLD   0 ETAS,   0 ELEMENTS
```

MAXIMUM ROW ERROR = 0.00000E+00   NEW   0 ETAS,   0 ELEMENTS

    0 STRUCTURAL COLUMNS

INVERT STATISTICS   [Begin solution of Subproblem 1 with

    4 NONZ IN BASIS   a basis factorization. The initial basis is

    0 STRUCTURAL COLUMNS IN BASIS   all logicals.]

    0 VECTORS ABOVE BUMP

    4 VECTORS BELOW BUMP

L   14 NONZ   0 ETAS

U   0 NONZ   0 ETAS

TOTALS   14 OFF DIAG NONZ   0 ETAS

SUBPROBLEM   1

ITCOUNT	STATUS	OBJ.VALUE	VECIN	VECOUT	DJ
	NETA	NELEM	TIME	NO.CAND	CRIT
1 M	F	-10.000000	5	4	-1.0000000
	0	0	0.00	0	
2	F	-14.000000	6	3	-0.50000000
	1	4	0.00	0	
2		-14.000000	6	3	0.0000E+00
	2	8	0.00	0	

[In iteration 1, the initial solution is feasible. The eta file is empty because of the initial all-logical basis. Column 5 with a reduced cost of -1.0 enters and column 4 leaves. The 'M' after the iteration count indicates a major iteration in multiple pricing, i.e. all nonbasic columns are priced. Iteration 2 is a minor iteration implying that one of the candidate columns in the previous pricing operation now enters the basis. This is column 6 which replaces column 3. Before this basis change, there is one eta with four nonzeros. Note that this information on the eta file always refers to the basis before the pivot. After the pivot, the value of the objective is -14.0. The next step shows that the solution is optimal, i.e. status = ' '. Note that iteration count, vecin and vecout are not updated. The optimal basis has two etas with eight nonzeros. Also, a proposal is generated automatically and not registered under "No.Cand".]

```
    NAME        SUB2                        [The above input and first solution proce-
    ROWS                                    dure is now repeated for Subproblem 2]
    COLUMNS
    RHS
    ENDATA
```

PROBLEM STATISTICS

    5 ROWS

    2 STRUCTURAL COLUMNS

    8 NON-ZERO ELEMENTS

DENSITY = 80.00 PERCENT

INPUT TIME SUBPROBLEM SUB2   =   0.00 SECONDS

   INVERSION SUBPROBLEM   2   OLD   0 ETAS,   0 ELEMENTS

   MAXIMUM ROW ERROR = 0.00000E+00   NEW   0 ETAS,   0 ELEMENTS

     0 STRUCTURAL COLUMNS

INVERT STATISTICS

   5 NONZ IN BASIS

   0 STRUCTURAL COLUMNS IN BASIS

   0 VECTORS ABOVE BUMP

   5 VECTORS BELOW BUMP

L   16 NONZ   0 ETAS

U   0 NONZ   0 ETAS

TOTALS   16 OFF DIAG NONZ   0 ETAS

SUBPROBLEM   2

ITCOUNT	STATUS	OBJ.VALUE	VECIN	VECOUT	DJ
NETA	NELEM	TIME	NO.CAND	CRIT	
1 M	F	-20.000000	6	3	-2.0000000
0	0	0.00	0		
2	F	-25.000000	7	5	-1.0000000
1	4	0.00	0		
2		-25.000000	7	5	0.00000E+00
2	8	0.00	0		

TOTAL INPUT TIME =   0.00   SECONDS

[At this point, both of the subproblems have been optimized for the first time. Each has generated a proposal for the master problem.]

BEGINNING MAJOR CYCLE   1

TIME FOR PREVIOUS CYCLE = 0.00 SEC. TIME IN MASTER = 0.00 SEC.

   INVERSION SUBPROBLEM   0   OLD   0 ETAS,   0 ELEMENTS
   MAXIMUM ROW ERROR =   0.35527E-14 NEW   2 ETAS,   6 ELEMENTS
      2 STRUCTURAL COLUMNS

INVERT STATISTICS   [Begin solution of master problem which is
   8 NONZ IN BASIS   designated as Subproblem 0.]
   2 STRUCTURAL COLUMNS IN BASIS
   0 VECTORS ABOVE BUMP
   4 VECTORS BELOW BUMP
L   0 NONZ   0 ETAS
U   6 NONZ   2 ETAS
TOTALS 4 OFF DIAG NONZ   2 ETAS

RESTRICTED MASTER PROBLEM [The master problem has two proposals.]

TOTAL NUMBER OF CANDIDATES 2 WITH 6 ELEMENTS

MEAN DENSITY OF CANDIDATES 75.00 PERCENT

ITCOUNT	STATUS	OBJ.VALUE	VECIN FROM	VECOUT FROM	DJ
NETA	NELEM	TIME			

MASTER DUAL VARIABLES

0.000000E+00 1.00000   −22.0000   −25.0000

[The dual prices of the coupling constraints are saved for use in the subproblems.]

SUBPROBLEM   1

ITCOUNT	STATUS	OBJ.VALUE	VECIN	VECOUT	DJ
NETA	NELEM	TIME	NO.CAND	CRIT	
3 M	F	−11.999999	3	6	−0.59999996
	2	8	0.00	1	R
4	F	−21.999985	4	5	−0.49999994
	3	12	0.00	2	R
4		−21.999985	4	5	0.0000E+00
	4	16	0.00	2	O

[Subproblem 1 is solved with the prices from the master problem. The iteration count is cumulative, hence we start with iteration 3. At the end of this iteration, a proposal is generated under the "percentage improvement" criterion. For an explanation of proposal generation criteria, look up QPROP in the Variable Dictionary in §5.5.3 and §3.3.6 of Subroutine CHECK.]

SUBPROBLEM   2

ITCOUNT	STATUS	OBJ.VALUE	VECIN	VECOUT	DJ
NETA	NELEM	TIME	NO.CAND	CRIT	
3 M	F	-4.9999991	3	4	-1.0000000
2	8	0.00	1	R	
4	F	-14.999999	5	6	-2.0000000
3	13	0.00	2	R	
5 M	F	-24.999985	4	7	-1.0000000
4	17	0.00	3	R	
5		-24.999985	4	7	0.00000E+00
5	21	0.00	3	O	

[Subproblem 2 has generated three proposals for the master problem in this cycle. Note that the proposal corresponding to the optimal solution has already been sent prior to satisfying the optimality criterion. Therefore it is not generated again.]

BEGINNING MAJOR CYCLE   2

TIME FOR PREVIOUS CYCLE = 0.00 SEC.   TIME IN MASTER = 0.00 SEC.

RESTRICTED MASTER PROBLEM

 TOTAL NUMBER OF CANDIDATES   7 WITH   19 ELEMENTS

 MEAN DENSITY OF CANDIDATES 67.86 PERCENT

ITCOUNT	STATUS	OBJ.VALUE	VECIN	FROM	VECOUT	FROM	DJ
NETA	NELEM	TIME					
1 M	I	6.9999991	13	2	8	2	-249.99999
2	6	0.00					
2	F	-23.999985	12	2	13	2	-7.0534554
3	9	0.00					

3 M	F	−31.999985	5	0	2	0	−1.0000000
4	12	0.00					
4 M	F	−39.999985	9	1	7	1	−5.6427641
5	14	0.00					
4		−39.999985	9	1	7	1	000000E+00
6	17	0.00					

MASTER DUAL VARIABLES

1.00000	1.00000	0.173834E−14	0.000000E+00

[The master problem is solved with seven proposal: two old ones; two new ones from Subproblem 1 and three new ones from Subproblem 2. In the first iteration, the solution is infeasible, i.e. status = 'I'. The Phase 1 objective (sum of infeasibilities) is 7.0. The infeasibilities are removed in the next iteration. The feasible solution has a Phase 2 objective of -24.0. At optimality, the objective value is -40.0. New prices are sent to the subproblems.]

SUBPROBLEM   1

ITCOUNT	STATUS	OBJ.VALUE	VECIN	VECOUT	DJ
NETA	NELEM	TIME	NO.CAND	CRIT	
4 M		0.40607352E−15	9	7	00000E+00
4	16	0.00	0	O	

SUBPROBLEM   2

ITCOUNT	STATUS	OBJ.VALUE	VECIN	VECOUT	DJ
NETA	NELEM	TIME	NO.CAND	CRIT	
5 M		0.00000000E+00	9	7	0.0000E+00

```
   5       21          0.00        0        O
```

[No new proposals are generated in this cycle.]

BEGINNING MAJOR CYCLE   3

TIME FOR PREVIOUS CYCLE = 0.00 SEC.   TIME IN MASTER = 0.00 SEC.

RESTRICTED MASTER PROBLEM

 TOTAL NUMBER OF CANDIDATES   7 WITH   19   ELEMENTS
 MEAN DENSITY OF CANDIDATES   67.86 PERCENT

ITCOUNT   STATUS   OBJ.VALUE   VECIN FROM VECOUT FROM   DJ
     NETA   NELEM   TIME
  LOWER BOUND   =   -40.00000000   [The lower bound on the primal objective]
  RELATIVE DIFFERENCE =   0.0000000   [Duality gap = 0 => Optimality]

PROPOSAL GENERATION STATISTICS :   [A summary of proposals generation.]
CRITERION   R   F   A   B   O   U
NO PROPOSALS   5   0   0   0   4   0

SUBPROBLEM MASTER   [Solution of the master problem]

JH(I)	VALUE	ROW NAMES	PI(I)	RHS
OBJ	40.000000	OBJ	1.0000000	0.000000E+00
XO	20.000000	RM1	1.0000000	40.000000
9 1	1.4177447	CROW 1	0.17383413E-14	1.0000000
12 2	1.4177447	CROW 2	0.00000E+00	1.0000000
EOF				

[The logical variable OBJ for the objective row is basic in its row. Its value is 40.0. Since this logical is the negative of the objective value, the latter is -40.0. See (2.11) in §2.2.3. The coupling variable X0 is basic in row RM1 with a value of 20.0. The other basic variables are proposals: column 9 which came from Subproblem 1 and column 12 which came from Subproblem 2. Since these are not original variables in the LP, they can be ignored. All nonbasic variables are of course at zero value. The dual variables for the constraints are listed under the PI(I) column.]

TRANSF TIME =   0.00 SECONDS

INPUT/OUTPUT TIME =   0.00   SECONDS

[Beginning of Phase 3 to reconstruct a primal solution to the subproblems.]


```
    INVERSION SUBPROBLEM    1            OLD  4 ETAS, 16 ELEMENTS
    MAXIMUM ROW ERROR = 0.00000E+00      NEW  0 ETAS,  0 ELEMENTS
       0 STRUCTURAL COLUMNS
INVERT STATISTICS
    4 NONZ IN BASIS
    0 STRUCTURAL COLUMNS IN BASIS
    0 VECTORS ABOVE BUMP
    4 VECTORS BELOW BUMP
L    14 NONZ    0 ETAS
U     0 NONZ    0 ETAS
TOTALS    14 OFF DIAG NONZ    0 ETAS


    INVERSION SUBPROBLEM    1            OLD  0 ETAS, 0 ELEMENTS
    MAXIMUM ROW ERROR =  0.00000E+00     NEW  0 ETAS, 0 ELEMENTS
       0 STRUCTURAL COLUMNS
INVERT STATISTICS
    4 NONZ IN BASIS
    0 STRUCTURAL COLUMNS IN BASIS
```

```
    0  VECTORS ABOVE BUMP

    4  VECTORS BELOW BUMP

L    14 NONZ     0 ETAS

U     0 NONZ     0 ETAS

TOTALS     14 OFF DIAG NONZ     0 ETAS
```

[Note that the first call to INVERT was in conjuction with the recovery of certain dual values. The second call marked the beginning of the solution procedure for the Phase 3 subproblem.]

SUBPROBLEM   1

ITCOUNT	STATUS	OBJ.VALUE	VECIN	VECOUT	DJ
	NETA	NELEM	TIME	NO.CAND	CRIT
5 M	I	9.9999991	6	2	−2.0000000
	0	0	0.00	0	
6	F	0.00000000E+00	5	4	−0.50000000
	1	4	0.00	0	
6		0.00000000E+00	5	4	0.000000E+00
	2	8	0.00	0	

SUBPROBLEM SUB1   [Solution of Subproblem 1]

JH(I)	VALUE	ROW NAMES	PI(I)	RHS
OBJ	0.000000E+00	OBJ	1.0000000	−10.000000
X2	−0.11102230E−14	RM1	1.0000000	10.000000
RS11	20.000000	RS11	0.00000E+00	30.000000
X1	10.000000	RS12	0.00000E+00	20.000000

EOF

[X2 is basic at 0.0 (i.e. degenerate solution). The logical (slack) for row RS11 is 20.0. X1 is 10.0.]

TRANSF TIME =   0.00 SECONDS

INPUT/OUTPUT TIME =   0.00   SECONDS

INVERSION SUBPROBLEM   2   OLD 5 ETAS, 21 ELEMENTS

MAXIMUM ROW ERROR = 0.00000E+00   NEW   0 ETAS,   0 ELEMENTS

    0 STRUCTURAL COLUMNS

INVERT STATISTICS

    5 NONZ IN BASIS

    0 STRUCTURAL COLUMNS IN BASIS

    0 VECTORS ABOVE BUMP

    5 VECTORS BELOW BUMP

L   16 NONZ   0 ETAS

U   0 NONZ   0 ETAS

TOTALS   16 OFF DIAG NONZ   0 ETAS

INVERSION SUBPROBLEM   2   OLD 0 ETAS,   0 ELEMENTS

MAXIMUM ROW ERROR = 0.00000E+00   NEW 0 ETAS,   0 ELEMENTS

    0 STRUCTURAL COLUMNS

INVERT STATISTICS

    5 NONZ IN BASIS

    0 STRUCTURAL COLUMNS IN BASIS

    0 VECTORS ABOVE BUMP

    5 VECTORS BELOW BUMP

L   16 NONZ   0 ETAS

U   0 NONZ   0 ETAS

TOTALS   16 OFF DIAG NONZ   0 ETAS

SUBPROBLEM   2   [Solution of Subproblem 2]

ITCOUNT	STATUS	OBJ.VALUE	VECIN	VECOUT	DJ
	NETA	NELEM	TIME	NO.CAND	CRIT
6 M	I	9.9999991	6	2	-2.0000000
	0	0	0.00	0	
6		0.00000000E+00	6	2	0.00000E+00
	1	4	0.00	0	

SUBPROBLEM SUB2

JH(I)	VALUE	ROW NAMES	PI(I)	RHS
OBJ	0.000000E+00	OBJ	1.0000000	-10.000000
Y1	5.0000000	RM1	1.0000000	10.000000
RS21	5.0000000	RS21	0.000000E+00	10.000000
RS22	10.000000	RS22	0.000000E+00	10.000000
RS23	10.000000	RS23	0.000000E+00	15.000000

EOF

[Y1 is 5.0. Y2 is nonbasic and hence 0. The logicals (slacks) are 5.0, 10.0 and 10.0 for rows RS21, RS22 and RS23 respectively.]

TRANSF TIME =   0.00 SECONDS

INPUT/OUTPUT TIME =   0.00   SECONDS

MAXIMUM ROW ERROR IN COMMON ROWS =   0.66613E-14

TOTAL TRANSFORMATION TIME =   0.00   SECONDS

TOTAL INPUT/OUTPUT TIME =   0.00   SECONDS

5.5 Variable Dictionary

The program has four different common storage areas : (a) COMMON BLOCK, (b) COMMON FT09, (c) COMMON INDISK and (d) COMMON blank. COMMON BLOCK contains problem parameters and other data that set the dimension of the arrays in the program. COMMON FT09 contains information that specifies what and where to write on the direct access file. COMMON INDISK contains the information of the problem that is to be swapped between the direct access file and core memory. COMMON blank contains other variables that will be used during the running of the program. The following is a list of variables defined in the COMMON statements. The type (I*2 for half-word integer, etc.) of the variable is given in parenthesis after its description.

5.5.1. Common BLOCK

KMULT : Number of candidates in multiple pricing. Maximum is five. (I*2)

LTRAC : Record length in bytes of the direct access file used to store master and subproblem data. Set to 13028. (I*2)

MAST : Control variable. Set to '0' in BLOCK DATA. Kept at '0' if there is no subproblem. Set to '1' in Phase 1 and Phase 2 of D-W algorithm (in INDATA). Set to '2' in Phase 3 (in RESULT). With NPH = 1 and MAST = 0, a subproblem is solved to optimality right after its data has been read. (I*2)

NCAND : Number of proposals generated in the current master major cycle that have not yet been packed. Set to zero in BLOCK DATA. Reset to zero after a pack operation. (I*2)

NEMAX : Maximum number of non-zero elements in data array (matrix and eta file) for a subproblem. The matrix file of the master problem can have up to LPOINR-1 non-zero elements. Its eta file can have up to (NEMAX - LPOINR) non-zero elements. Set to 10,000. (I*2)

NCRMAX : Maximum number of coupling rows. Set to 99. (I*2)

NLAMAX : Maximum number of columns in master or subproblem, including logicals. Set to 1000. (I*2)

NPROS : Maximum number of proposals in the proposal buffer. Set to 25. (I*2)

NRMAX : Maximum number of rows in a problem. Set to 400. (I*2)

NTMAX : Maximum number of transforms (eta vectors) in the eta file. Set to 800. (I*2)

NZMAX : Maximum number of entries in a proposal. It is equal to the maximum number of coupling rows (including objective) plus one. Set to 100. (I*2)

QQ   :   Character string 'CROW' used to form convexity row names. (I*4)

Qx   :   Character Q concatenated by x where x stands for one of the following:, A, B C, E, F, H, I, L, N, O, R, U, M, and blank. Used to define various character constants.

ZTCOST   :   Zero tolerance in reduced cost. Set to 1.0E-04. (R*8)

ZTOLPV   :   Pivot tolerance. Set to 1.0E-06. (R*8)

ZTOLRJ   :   Feasibility tolerance used in CHUZR, FORMC, and INVERT. Set to 1.0E-07. (R*8)

ZTOLZE   :   Zero tolerance used for data and computations. Set to 1.0E-07. (R*8)

ZTCAND   :   Tolerance used in proposal generation. Also used to determine which coupling constraint must be enforced in Phase 3 of D-W algorithm. Set to 1.0E-02. (R*8)

5.5.2. *Common FT09*

NOLA(*, *):   Control array. Let L be the subproblem index. For the master, L= LMAX+1. NOLA(2, L) equals LE(NETA+1) - 1 and points to end of the eta file. NOLA(3, L) points to secondary storage of data. NOLA(4, L) points to secondary storage of rows and columns names. NOLA(1, LMAX+1) points to secondary storage of restart information. Dimension is NOLA(4, 11).

KD   :   Control variable associated with direct access file 9. (I*4)

LMAXP1   :   Number of subproblems plus one. (I*2)

5.5.3. *Common INDISK*

A(*)   :   Data array. For a subproblem, the eta file starts right after the matrix data. For the master, A(1) to A(LPOINR-1) contain the matrix data (including proposals). The eta file starts from A(LPOINR) onward. This is to allow room for the addition of proposal data. Dimension is A(10000). (R*8)

IA(*)   :   Row index of the *th element of the matrix or eta file. Dimension is IA(NEMAX). (I*2)

ISTYPE(*) : Indicates the type of the $*^{th}$ row. '0' for non-binding row, '1' for inequality row and '-1' for equality row.
Dimension is ISTYPE(NRMAX). (I*2)

ITCNT : Simplex iteration counter. (I*2)

ITSINV : Number of Simplex iterations since last invert. (I*2)

JH(*) : Index of column that is basic in the $*^{th}$ row.
Dimension is JH(NRMAX). (I*2)

KINBAS(*): Indicates the status of the $*^{th}$ column in the current solution. '0' if column is non-basic and 'i ' if column is basic in row i.
Dimension is KINBAS(NLAMAX). (I*2)

LA(*) : Pointer to the $*^{th}$ column in A(*). Dimension is LA(NLAMAX). (I*2)

LE(*) : Pointer to the $*^{th}$ eta vector in A(*). Dimension is LE(NTMAX+1). (I*2)

NCOL : Number of columns in the problem (logicals, structurals and rhs). (I*2)

NELEM : Number of non-zero elements in matrix file. (I*2)

NETA : Number of eta vectors currently in the eta file. When NETA is zero it triggers a call to INVERT. (I*2)

NROW : Number of rows in the problem being solved. (I*2)

NT : Cumulative transformation time in BTRAN and FTRAN. (I*4)

NUMCAN : Number of proposals generated during last cycle. (I*2)

NX : Cumulative I/O time. (I*2)

RHSCOL : Index of rhs column in use. (I*2)

ITIM : Start time for NORMAL in a cycle. (I*4)

ITINV : Auxilliary variable. (I*4)

JTIM : Auxiliary variable. (I*4)

JTINV : Cumulative transformation time. (I*4)

5.5.4. *Common Blank*

COND : Approximate condition number of the basis inverse. Computed in CHSOL. (R*4)

DCMIN(*) : Reduced cost of the $*^{th}$ column selected in a multiple pricing pass.
Dimension is DCMIN(5). (R*8)

DE : Value of the column leaving the basis. Also, after CHUZR, value at which a candidate column would enter the basis if pivoted in. Used also as auxiliary variable. (R*8)

DEOB(*) : Value of the objective value minus dual variable of the corresponding convexity constraint of the *th subproblem after treatment in NORMAL. Dimension is DEOB(LMAXP1). (R*8)

DNEW : New value of dual objective function. It is computed in MASTER at the beginning of a cycle when all subproblems have been solved to optimality. (R*8)

DOB : Objective value minus dual variable of convexity row for subproblem currently treated by NORMAL. (R*8)

DOBMA : Auxiliary variable used in CHECK. (R*8)

DOLD : Old value of dual objective value when starting a new cycle in MASTER. (R*8)

DP : Value of variable entering the basis. Computed in UPBETA. Tested in CHECK. Also used as auxiliary variable. (R*8)

DPROD : Auxiliary variable. (R*8)

DSUM : Auxiliary variable. (R*8)

DY : Potential improvement in objective value after calling CHUZR with JENTRY = 2. Also used as auxiliary variable. (R*8)

ERMAX : Maximum row error in absolute value. Computed in CHSOL. (R*8)

ICAND : When the proposal buffer is only partially filled at the end of one subproblem, ICAND marks the last proposal from this subproblem so that it can be followed by the first proposal from the next subproblem. (I*2)

ICASUB(*): Number of proposals generated at completion of *th subproblem. Dimension is ICASUB(10). (I*2)

ICLAST : Cycle at which the algorithm is to be stopped. (I*2)

ICOLS(*) : Array used for master problem only. Indicates the subproblem origin of the *th proposal. Dimension is ICOLS(NLAMAX). (I*2)

ICPI : Cycle at which a special set of prices is to be read and used instead of the master dual vector. (I*2)

ICUMB(*) : Column index of the *th extreme ray generated when solving a subproblem. Dimension is ICUMB(8). (I*2)

IFFEZ : '0' if the problem on hand is currently infeasible; '1' otherwise. (I*2)

INCO : Not used in present version of code.

INVFMA : Master problem inversion frequency. (I*2)

INVRQ : Subproblem inversion frequency. (I*2)

IOBJ : Index of the objective row. (I*2)

IPROP(*) : IPROP(I), I=1,..., 6, stores the number of proposals generated under the criteria (QPROP=) R (percentage improvement), F (frequency), A (Test A), B (Test B), O (optimality), and U (unboundedness), respectively. See § 3.3.6 as well as QPROP. Dimension is IPROP(6). (I*2)

IPROS : Pointer to first free position in the proposal buffer. (I*2)

IROWP : Pivot row index. (I*2)

IRST : Restart parameter. See § 5.2. (I*2)

ISAVE : '1' indicates that restart information is to be saved. See § 5.2. (I*2)

ISOL : '1' indicates that a Phase 3 is to be initiated after ICLAST cycles or at optimality. (I*2)

ISUNB : When non-zero indicates that the problem is unbounded and the number of extreme rays that have actually been generated. (I*2)

ITMA : Set to ITCNT when entering NORMAL. (I*2)

ITSLCA : Iteration counter since generation of last extreme point proposal. Set to '-5' when entering NORMAL to treat a subproblem. (I*2)

IVIN : Index of column entering the basis. (I*2)

IVOUT : Index of column leaving the basis. (I*2)

IYIV : Auxiliary index. (I*2)

JCOLP(*) : Index of the $*^{th}$ column selected as candidates in a multiple pricing pass. Dimension is JCOLP(5). (I*2)

KCYC : Cycle counter for the D-W algorithm. Equivalent to MAJOR in MASTER. (I*2)

KFASE : Indicates where in the buffer YA(*, *) the cost vector is to be formed. '1' indicates that the current solution of the subproblem on hand is infeasible. In this case the cost vector is formed in YA(*, 5). Otherwise, it is formed in YA(*, 6). (I*2).

KINP : Unit number of device on which master and subproblem data is read. (I*4)

KRIT(*) : Indicates stopping criterion when solving subproblem LS(*). '1' means no proposal is generated until optimality or unboundedness. '2' means up to NC(*) proposals can be generated. Dimension is KRIT(10). (I*2)

KSTR : Control parameter for computational strategies. See § 5.2. (I*2)

LCORE : Indicates which subproblem is currently in core. '0' implies master. (I*4)

LMAX : Number of subproblems. (I*2)

LS(*) : Index of the $*^{th}$ subproblem to be treated in the current cycle.

		Master problem is treated when LS(*) is zero. Dimension is LS(11). (I*2)
LSUB	:	Index of subproblem being processed. '0' implies master. (I*2)
MAXCA	:	Maximum number of proposals allowed (kept when KSTR is '5') when treating a subproblem. (I*2)
MAXNCA	:	Maximum number of proposals allowed (kept when KSTR is '5') in a master cycle. (I*2)
MSTAT	:	Current status of the problem being solved. (I*2)
MSTATU(*):		Status of the $*^{th}$ subproblem after its treatment. MSTATU(LMAXP1) gives the status of the master. Status is 'I' for currently infeasible solution, 'N' for no feasible solution, ' ' for optimal solution and 'U' for unbounded solution. (I*2)
MULT	:	Number of candidates selected in a multiple pricing pass. (I*2)
NAME(*)	:	Array of proposal labels. Used to form proposal labels stored in array ICOLS. NAME(*) contains the subproblem index if the $*^{th}$ proposal is an extreme point generated by the subproblem. It contains the negative of the subproblem index if the proposal is an extreme ray. (I*4)
NC(*)	:	Number of proposals to be generated (kept when KSTR is '5') when treating subproblem LS(*).
NCANFR	:	Proposal generation frequency. See § 5.2. (I*2)
NCASUB	:	Number of proposals generated in subproblem being solved. (I*2)
NCOLO	:	Number of columns in the master excluding proposal columns. (I*2)
NROWO	:	Number of coupling rows. (I*2)
NCUMB(*):		Column index of the $*^{th}$ extreme ray generated when treating a subproblem. The corresponding column is temporarily marked as being fixed at zero value. Dimension is NCUMB(5). (I*2)
NLETA	:	Number of L etas in the L-U factorization of the basis. (I*2).
NTEMP(*) :		Auxiliary array used mainly in input. Dimension is NTEMP(10). (I*2)
QPROP	:	Indicator of proposal generation criterion:

QPROP (continued):

' ' = no proposal;

'R' = percentage of improvement;

'F' = frequency;

'A' = test A;

'B' = test B;

'O' = optimality;

'U' = unboundedness.

(I*2)

RCTEST	:	Test value used for proposal generation based on reduced cost. Set to $(1.0E+4)*ZTCOST$. (R*4)
RPERC	:	Minimum percentage of relative improvement in subproblem objective function to generate a proposal according to "percentage of improvement" criterion. Set to 40 %. (R*4)
RSTOP	:	Convergence tolerance given by relative gap between current primal and best dual solution (in %). Set to .01 %. (R*4)
SUMINF	:	Sum of infeasibilities in a problem. (R*4)
TIMER	:	Time information for printing purposes. (R*4)
USERHS	:	Index of rhs in use. Default is 1. (I*2)
X(*)	:	Contains the current basic solution. Dimension is X(NRMAX). (R*8)
YA(*, *)	:	Working storage. Vectors YA(*, 1) to YA(*, 5) are used to expand (unpack) the columns selected for multiple pricing and there can be at most KMULT of these. YA(1, 6) to YA(NROWO, 6) contains the negative of the dual prices of the master problem. YA(*, 6) is then used to compute the simplex multipliers of the subproblem. When KFASE is '1', YA(*, 5) is used to compute the simplex multipliers corresponding to the infeasibility form for a subproblem. Dimension is YA(NRMAX, 6). (R*8)
YTEMP(*)	:	Negative of the simplex multiplier of the $*^{th}$ convexity row. Dimension is YTEMP(10). (R*8)
Z(*, *)	:	Proposal buffer. Dimension is Z(NZMAX, NPROS). (R*8)

Bibliography

R.M. Burton and B. Obel, "The multilevel approach to organizational issues of the firm - a critical review," *OMEGA Int. J. of Mgmt. Sci.* 5 (1977) 395-414.

R.M. Burton and B. Obel, "The efficiency of the price, budget and mixed approaches under varying a priori information levels for decentralized planning," *Management Science* 26 (1980) 401-417.

J. Christensen and B. Obel, "Simulation of decentralized planning in two Danish organizations using linear programming decomposition," *Management Science* 26 (1980) 401-417.

V. Chvátal, *Linear Programming*, W. H. Freeman, New York, (1983).

D. Dewitt, R. Finkel and M. Solomon, "The CRYSTAL multicomputer: design and implementation experience," *IEEE Transaction on Software Engineering 8* (1987) 953-966.

G.B. Dantzig, *Linear Programming and extensions*, Princeton University Press, Princeton, (1963).

G.B. Dantzig, M.A.H. Dempster and M.J. Kallio, (eds.), *Large-Scale Linear Programming, Vols 1 & 2*, IIASA, Laxenberg, Austria, 1981.

G.B. Dantzig and P. Wolfe, "The decomposition principle for linear programs," *Operations Research* 8 (1960) 101-111.

Y. Dirickx and P. Jennergren, *Systems Analysis by Multilevel Methods: with Applications to Economics and Management*, Wiley, New York, 1979.

Electric Power Research Institute (EPRI), "Decomposition of linear programs using concurrent processing on multicomputers," RP199-11 Final Report, 1989.

J.J.H. Forrest and J. A. Tomlin, "Updating Triangular Factors of the Basis to Maintain Sparsity in Product Form Simplex Method," *Mathematical Programming* 2 (1972) 263-278.

L.M. Goreux and A.S. Manne (eds.), Multilevel Planning: Case Studies in Mexico, North-Holland, Amsterdam, 1973.

H. Greengerg, (ed.), *Design and Implementation of Optimization Software*, Sijthoff & Noordhoff, Netherlands, 1978.

L. Haynes, R. Lau, D. Siewiorek and Mizell, "A survey of highly parallel computing," *Computer* 1 (1982) 9-24.

E. Hellerman and D. Rarick, "Reinversion with the preassigned pivot procedure," *Mathematical Programming* 1 (1971) 195-216.

W. Hillis, "Why parallel processing is inevitable?" in Kirkland and Poore (eds.), *Supercomputers*, Praeger, New York, 1987.

J.K. Ho, "Recent advances in the decomposition approach to linear programming," *Mathematical Programming Study* 31 (1987) 119-127.

J.K. Ho and S.K. Gnanendran, "Distributed decomposition of block-angular linear programs on a Hypercube computer," Technical Report, Management Science Program, University of Tennessee, Knoxville, TN 37996 (1989).

J.K. Ho and T.C. Lee, "Dynamics of information in distributed decision systems," Technical Report, Management Science Program, University of Tennessee, Knoxville, TN 37996 (1989).

J.K. Ho, T.C. Lee and R.P. Sundarraj, "Decomposition of linear programs using parallel computations," *Mathematical Programming* 42 (1988) 391-405.

J.K. Ho and E. Loute, "An advanced implementation of the Dantzig-Wolfe decomposition algorithm for linear programming," *Mathematical Programming* 20 (1981) 303-326.

J.K. Ho and E. Loute, "Computational experience with advanced implementation of decomposition Algorithms for linear programming," *Mathematical Programming* 27 (1981) 282-290.

J.K. Ho and E. Loute, "Computational aspects of DYNAMICO: a model of trade and development in the world economy," *R.A.I.R.O. Recherche opérationelle / Operations Research* 18 (1984) 403-414.

J.K. Ho and W. McKenney, "Triangularity of the basis in linear programs for material requirements planning," *Operations Research Letters* 7 (1988) 273-278.

IBM, *IBM Mathematical Programming System Extended/370 (MPSX/370) Program Reference Manual*, SH19-1095-02, (1979).

L.S. Lasdon, *Optimization Theory for Large Systems*, McMillan, London, (1970).

B.A. Murtagh, *Advanced Linear Programming : Computation and Practice*, McGraw Hill, New York, (1981).

B.A. Murtagh and M.A. Saunders, "Large-scale linearly constrained optimization," *Mathematical Programming* 14 (1978) 41-72.

L. Nazareth, "A land management model using Dantzig-Wolfe decomposition," *Management Science* 26 (1980) 510-523.

W. Orchard-Hays, *Advanced Linear Programming Computing Techniques*, McGraw Hill, New York, (1968).

J.E. Samouilidis and A. Arabatzi-Ladia, "Modeling decentralized decision making in the energy sector," *OMEGA Int. J. of Mgmt. Sci.* 12 (1984) 437-447.

J.A. Tomlin, "LPM1 User's Guide," Unpublished manuscript, Systems Optimization Laboratory, Stanford University, 1973.

H.P. William and A. Redwood, "A structured program model in the food industry," Operational Research Quarterly 25 (1974) 517-527.

Index

accuracy 49

allocation 25

basis 5

    augmented 8

    factorization 7, 91

    inverse 9

    factorization 91

block 22

bump 94

constraints, block 23

    convexity 24

    coupling 23

    non-binding 8

core 36

cycle 23

    major 121

    partial 133

decomposition, parallel 2

    principle 19

dimensions 172

direct access device 73

eta, file 8

    matrix 8

    vector 8

extreme point 19, 27

extreme ray 19, 28

factorization - LU 91

feasibility 24

files 172

fill-ins 95

format 175

Gaussian elimination 96

heuristics 96

infinity norm 120

input 173

linear program 5

    block-angular 22

    standard form 5

logicals 5

master problem 24

matrix, basic 5

    sparse 1

merit count 96

multicomputer 2

output 175

parameters 173

permutation 91

pivot 34

    sequence 92

portability 168

pricing multiple 10

print-out 182

problem, infeasible 24

    statistics 85, 176

    unbounded 25

proposals 22

    companion 40

    intermediate 41

    scaling 136

purging 132

ratios 55

reconstruction 25

residual 49

resource 150

revised simplex method 5

row error 52

singletons 91

statistics 175

strategies 28

subproblem 21

timer	173
tolerances	173
transformation	34, 67
unpacking	6
variables, artificial	5
logical	5
slack	5
structural	64
surplus	5
vector, allocation	149
cost	5
dual	5
eta	8
solution	5